REVISE AQA GCSE (9–1)
Physics

REVISION WORKBOOK

Higher

Series Consultant: Harry Smith

Author: Catherine Wilson

Also available to support your revision:

Revise GCSE Study Skills Guide   9781447967071

The **Revise GCSE Study Skills Guide** is full of tried-and-trusted hints and tips for how to learn more effectively. It gives you techniques to help you achieve your best – throughout your GCSE studies and beyond!

Revise GCSE Revision Planner   9781447967828

The **Revise GCSE Revision Planner** helps you to plan and organise your time, step-by-step, throughout your GCSE revision. Use this book and wall chart to mastermind your revision.

> For the full range of Pearson revision titles across KS2, KS3, GCSE, Functional Skills, AS/A Level and BTEC visit:
> www.pearsonschools.co.uk/revise

Contents

- - - - - - - - - - - - -

AQA publishes Sample Assessment Material and the Specification on its website. This is the official content and this book should be used in conjunction with it. The questions have been written to help you practise every topic in the book. Remember: the real exam questions may not look like this.

Energy stores and systems

1 Identify the correct energy stores that occur in the following examples by drawing lines to show your answers. The first one has been done for you.

A	A container box lifted by a crane
B	Hot water in a saucepan
C	Bag of coal
D	Moving wind turbine

chemical
gravitational
kinetic
thermal

(1 mark)

2 (a) Explain what is meant by a closed system.

A closed system is an isolated system where ....

.... **(1 mark)**

(b) Explain how your answer to (a) relates to the principle of the conservation of energy.

The total energy in a closed system is the same after the transfer ....

....

.... **(2 marks)**

3 The principle of the conservation of energy states that energy can be usefully transferred to other stores. Complete an energy flow diagram to show the changes in energy stores and transfers for a battery-operated music system, when it is switched on. Write the correct description of energy in each box.

> Your answer should include the following terms: 'electrical current', 'heating', 'chemical energy', 'sound'.

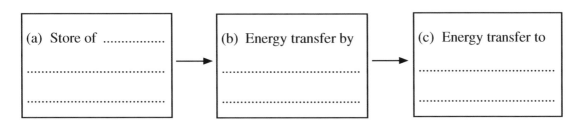

(a) Store of ....

....

....

(b) Energy transfer by

....

....

(c) Energy transfer to

....

....

(3 marks)

4 A basket of apples is lowered to the ground from an apple tree using a rope and pulley. The total energy transferred is 250 J. Identify the energy transfers taking place, including the useful and wasteful energy components that result.

....

....

....

....

....

....

.... **(4 marks)**

Changes in energy

1 Which is the correct equation for calculating gravitational potential energy? Tick **one** box.

> Always answer multiple-choice questions, even if you don't actually know the answer.

   ☐ $E_p = m\,v\,h$ ☐ $E_p = m\,F\,a$

   ☐ $E_p = \frac{1}{2}\,m\,v^2$ ☐ $E_p = m\,g\,h$ **(1 mark)**

Maths skills

Guided

2 A cyclist and her bicycle are travelling at 6 m/s.

The mass of the cyclist and bicycle is 70 kg.

Calculate the kinetic energy of the cyclist. Choose the correct unit from the box.

m/s^2	J	W

Kinetic energy = ...

so ....

                    Kinetic energy = ... unit .... **(3 marks)**

3 A spring has a spring constant of 200 N/m.

It is stretched 15 cm when a mass is applied.

Calculate the energy transferred to the spring. Use the correct equation from the Physics Equation Sheet.

....

....

               Energy transferred = ... J   **(3 marks)**

4 A dancer with a weight of 600 N practises chin-ups during training and raises her body 70 cm.

Calculate the gravitational potential energy gained by the dancer between the bottom and top of the chin-up.

> Remember to convert from weight to mass first!

....

....

           Gravitational potential energy = ... J   **(3 marks)**

Energy changes in systems

1 Give the equation for specific heat capacity.

.... **(1 mark)**

2 Calculate how much energy is needed to heat 0.8 kg of water from 30 °C to 80 °C. The specific heat capacity of water is 4200 J/kg °C. Use the correct equation from the Physics Equation Sheet.

....

....

....

Energy required = ... J **(3 marks)**

3 A 1.2 kg block of copper is supplied with 20 000 J of electrical energy. Calculate the change in temperature of the copper. The specific heat capacity of copper is 385 J/kg °C. Use the correct equation from the Physics Equation Sheet.

> You will need to rearrange the equation you used in Question 2.

Guided

$\Delta\theta = \Delta E \ / \ (m \ c)$, so $\Delta\theta = 20\ 000 \ /$ ....

....

....

Change in temperature of the copper = ... °C **(3 marks)**

4 A 0.8 kg block of metal is heated for 540 seconds with an electrical power of 30 W. The temperature increase is 25 °C. Calculate the specific heat capacity of the block of metal. Use the correct equation from the Physics Equation Sheet.

> You need to recall that $E = P\,t$ to calculate the electrical energy in, before calculating the energy out (energy supplied to the metal block).

thermometer

V

A

low voltage supply

metal block

lagging

heater

....

....

....

....

Specific heat capacity = ... J/kg °C **(4 marks)**

Specific heat capacity

1   Water is widely used in cooling systems because of its relatively high specific heat capacity compared with some other liquids.

   (a)   Write the definition of the term specific heat capacity.

   .... **(1 mark)**

   (b)   Give the three quantities that need to be measured to calculate the specific heat capacity of a substance.

   .... **(1 mark)**

2   (a)   Describe an experiment that could be set up to measure the specific heat capacity of water using an electric water heater, a beaker and a thermometer.

   > Remember 'pre-experiment' steps, e.g. zero the balance to eliminate the mass of apparatus before measuring substances, take a starting temperature reading before heating and decide on the range or type of measurements to be taken.

   ....

   ....

   ....

   ....

   ....

   ....

   ....

   .... **(5 marks)**

   (b)   Suggest how you can determine the amount of thermal energy supplied to the heater by the electric current.

   ....

   ....

   .... **(2 marks)**

   (c)   Explain how this experiment could be improved to give more accurate results.

   ....

   ....

   .... **(2 marks)**

3   A known mass of ice is heated until it becomes steam. The temperature is recorded every minute. Describe how to use the data to identify when there are changes of state.

   ....

   ....

   .... **(2 marks)**

Power

1 A kettle transfers 12 500 J of electrical energy in 5 seconds. What is the power rating of the kettle? Tick **one** box.

   ☐   2500 W

   ☐   12 500 W

   ☐   25 000 W

   ☐   62 500 W   **(1 mark)**

2 A microwave heats a drink in 20 seconds using 15 000 J of electrical energy. Calculate the power of the microwave.

Guided

Energy transferred = ... J, time taken = ... s

$P = E / t$ = ... W

                                                 Power = ... W   **(2 marks)**

3 A student with a mass of 60 kg climbs 20 stairs to a physics lab. Each stair is 0.08 m high; g is 10 N/kg.

   (a) Calculate the gravitational potential energy gained in climbing the 20 stairs to the lab. Select the correct unit from the box.

watts	newtons	joules

> You need to remember that $E_g = m\,g\,h$ because it's not on the Physics Equation Sheet.

....

....

                         Gravitational potential energy = ... unit ....   **(3 marks)**

   (b) Calculate the power of the student's muscles to climb the stairs in 12 seconds. Include the unit in your answer.

                                        Power = ...   **(1 mark)**

4 Altaf and Cathy investigate the time taken for two different winch motors to transfer energy.

> You will need to recall the equation $P = E / t$

   (a) Calculate the time taken for a 3 W winch motor to transfer 360 J of energy.

....

.... **(2 marks)**

   (b) Calculate the time taken for a 5 W winch motor to transfer 360 J of energy.

....

.... **(2 marks)**

Energy transfers and efficiency

1 (a) Identify the most suitable material, from the table below, for building an energy efficient garage.

Material	Relative thermal conductivity
brick	1.06
concrete	1.00
sandstone	2.20
granite	2.75

.... **(1 mark)**

(b) Define the term 'low relative thermal conductivity' of a material.

....

.... **(2 marks)**

2 (a) Some houses are built with very thick walls. Explain how these walls help to keep the houses warm in the winter.

Thicker walls provide more material for ....

to travel through from the inside to outside, so the ....

is less, keeping the houses warmer.   **(2 marks)**

(b) In hot countries, some traditional houses have thick walls with small windows. Explain why.

Thicker walls provide more material for ....

to travel through from the outside to inside, so the ....

is less, keeping the house cool.   **(2 marks)**

3 A box gains 100 J of energy in its gravitational potential energy store when it is lifted from the floor to a lab desk. The motor lifting the box transfers 400 J as kinetic energy.

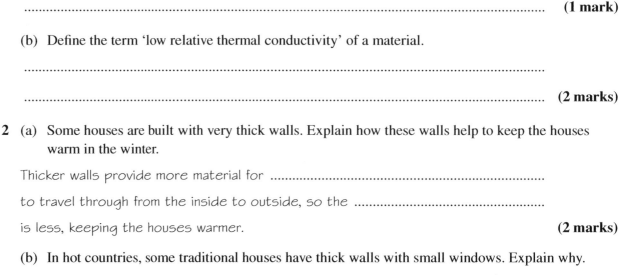

Remember: efficiency = $\dfrac{\text{useful energy transferred by the machine}}{\text{total energy supplied to the machine}}$

(a) Calculate the efficiency of the motor.

....

.... **(2 marks)**

(b) Suggest how the efficiency of the motor could be improved by reducing friction.

....

.... **(2 marks)**

4 A student uses four beakers containing hot water, each wrapped with different insulating materials, to investigate the transfer of thermal energy. Which factor will **not** affect the rate of transfer of thermal energy? Tick **one** box.

☐ Rate of data collection   ☐ Temperature of the room

☐ Starting temperature of water   ☐ Thickness of the insulators   **(1 mark)**

Thermal insulation

1 (a) Describe an experimental method, using the apparatus in the diagram below, to investigate the insulating properties of different materials.

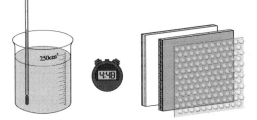

....

....

....

....

....

....

.... **(5 marks)**

(b) Identify the independent and dependent variables in this experiment.

....

....

.... **(2 marks)**

(c) Name **four** control variables in this experiment.

....

....

.... **(2 marks)**

2 Identify a hazard with conducting this experiment and explain how the risks to those conducting the experiment may be minimised.

....

.... **(1 mark)**

3 Suggest an alternative piece of apparatus **not** shown in the diagram that could be used for data collection to improve accuracy.

.... **(1 mark)**

4 Describe the conclusion that you would expect to reach at the end of the experiment which relates the independent and dependent variables.

....

....

.... **(2 marks)**

Energy resources

1 A hydroelectric power station is used to produce electricity when demand is high.

(a) Explain why the hydroelectric power station is a reliable producer of electricity.

> Hydroelectric power relies on water power; you need to explain why this is reliable.

....

....

....

.... **(2 marks)**

(b) Give **one** reason why we cannot use hydroelectric power stations in more places in the UK.

....

.... **(1 mark)**

2 For each of the following statements about fossil fuels, describe the negative environmental impacts.

> Think about the possible consequences of the statements describing the use of fossil fuels.

(a) Carbon dioxide is released as a result of burning fossil fuels.

....

....

....

.... **(2 marks)**

(b) Burning fossil fuels produces sulfur dioxide and nitrogen oxides.

....

....

....

.... **(2 marks)**

(c) Fossil fuels need to be extracted from the ground and transported to the power station.

....

....

....

.... **(2 marks)**

Patterns of energy use

1 The graphs show patterns of energy use and human population growth.

(a) Give **three** reasons why energy consumption rose significantly after the year 1900.

1. After 1900, the world's energy demand .... as the population grew.

2. There was development in ....

3. The rise of power stations using fossil fuels added to .... **(3 marks)**

(b) (i) Identify the **three** main energy resources used to provide the world's energy between the years 1000 and 2000.

.... **(1 mark)**

(ii) Suggest **two** reasons why the use of energy resources has increased in the developed world.

1. ....

2. .... **(2 marks)**

(iii) Give a reason why nuclear energy resources appear only after 1950.

.... **(1 mark)**

(iv) Identify a renewable resource from the graph that makes use of gravitational potential energy.

.... **(1 mark)**

2 In the graph in Q1, the patterns in energy consumption are similar to the pattern in the world's population growth. Describe the issues that may result from the continuing use of energy in the future, at the same rate as shown in the graph.

> Consider the finite resources and increasing demand.

....

....

....

....

.... **(4 marks)**

Extended response – Energy

Millie plays on a swing in the park. The swing seat is initially pulled back by her friend to 30° to the vertical position and then released. Describe the energy changes in the motion of the swing.

Your answer should also explain, in terms of energy, why the swing eventually stops.

> You will be more successful in extended response questions if you plan your answer before you start writing.
>
> The question asks you to give a detailed explanation of the energy changes as the swing moves backwards and forwards. Think about:
>
> - How gravitational potential energy changes as the swing is pulled back.
> - Points at where gravitational potential energy (E_p) and kinetic energy (E_k) are at maximum and at 0.
> - Where some energy may be lost from the system.
> - Why the swing will eventually stop.
>
> You should try to use the information given in the question.

....

....

....

....

....

....

....

....

....

....

....

....

....

....

.... **(6 marks)**

Circuit symbols

1   Select the component that is designed to respond to changes in levels of light. Tick **one** box.

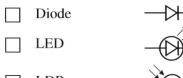

☐   Diode

☐   LED

☐   LDR

☐   thermistor

(1 mark)

2   (a)   The three symbols below represent three components. Write the name of each component in the corresponding box.

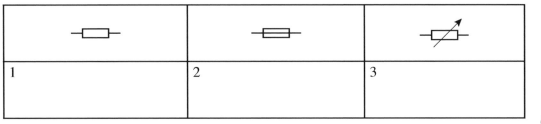

1	2	3

(3 marks)

   (b)   Identify which component in (a) is commonly found in a household plug connected to the live wire.

....   **(1 mark)**

3   Complete the table of circuit symbols below:

Component	Symbol	Purpose
ammeter		
		provides a fixed resistance to the flow of current
	—▷|—	
		allows the current to be switched on / off

(4 marks)

4   Draw a circuit that could be used to measure the resistance of an unknown resistor.

> Consider whether each component should be connected in series or in parallel.

....

....

....

....

....

....   **(4 marks)**

Electrical charge and current

1   The electric current flowing in a circuit is 4 A.

(a) Explain what is meant by an electric current.

....

.... **(2 marks)**

(b) The current flows for 8 seconds. Calculate how much charge has flowed.

Choose the correct unit from the box.

> You need to recall the equation: charge flow = current × time ($Q = I\,t$)

C	J	A

....

....

Charge = ... unit.................. **(3 marks)**

2   The diagram shows a series circuit.

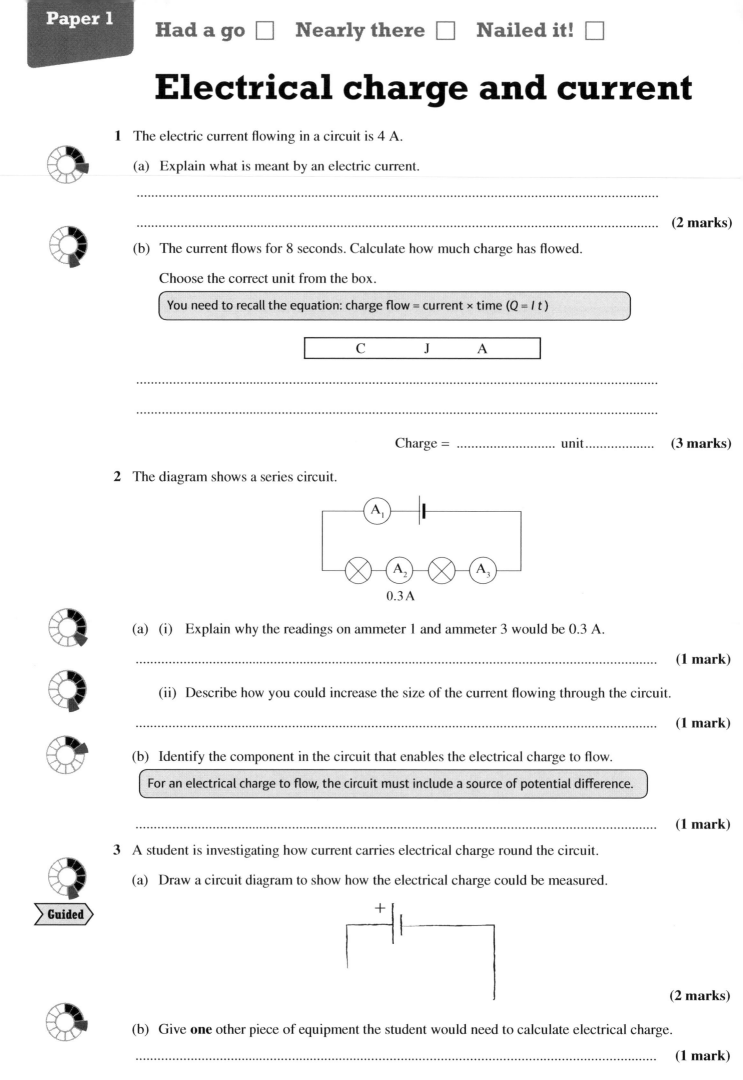

0.3 A

(a) (i)   Explain why the readings on ammeter 1 and ammeter 3 would be 0.3 A.

.... **(1 mark)**

(ii) Describe how you could increase the size of the current flowing through the circuit.

.... **(1 mark)**

(b) Identify the component in the circuit that enables the electrical charge to flow.

> For an electrical charge to flow, the circuit must include a source of potential difference.

.... **(1 mark)**

3   A student is investigating how current carries electrical charge round the circuit.

(a) Draw a circuit diagram to show how the electrical charge could be measured.

Guided

+

(2 marks)

(b) Give **one** other piece of equipment the student would need to calculate electrical charge.

.... **(1 mark)**

Current, resistance and pd

1   Which quantity is the ohm (Ω) the unit of? Tick **one** box.

☐ Current

☐ Energy

☐ Potential difference

☐ Resistance   **(1 mark)**

2   Give Ohm's law.

The current flowing through a .... at constant temperature is

.... to the potential difference across a resistor.   **(2 marks)**

3   Use Ohm's law to calculate the following:

> You will need to recall the equation: $R = V / I$

(a)   The resistance of a resistor with a potential difference of 12 V across it and a current of 0.20 A passing through it.

....

....

....

Resistance = ... Ω   **(2 marks)**

(b)   The current passing through a 55 Ω resistor with a potential difference of 22 V across it.

....

....

....

Current = ... A   **(2 marks)**

4   (a)   Sketch two lines on the graph to show two ohmic conductors of different resistances. Label these A and B.

(2 marks)

(b)   From your graph, identify which line represents the resistor with the higher resistance.

....

....

....

.... **(1 mark)**

Current in A

0   Potential difference in V

Practical skills

Investigating resistance

1 Which of these is the correct method of connecting an ammeter and a voltmeter to determine resistance of a component in a circuit? Tick **one** box.

☐ Ammeter and voltmeter are both connected in series with the component.

☐ Ammeter is connected in series but voltmeter is connected in parallel across the component.

☐ Ammeter and voltmeter are both connected in parallel across the component.

☐ Voltmeter is connected in series but ammeter is connected in parallel across the component.

(1 mark)

2 Calculate the resistance of a lamp supplied with a current of 1.5 A and a potential difference of 90 V.

> You need to recall the equation: $R = V / I$

....

....

....

Resistance of lamp = ... Ω   **(2 marks)**

3 Complete the circuit diagram to show how the resistance of a lamp may be obtained using an ammeter and a voltmeter.

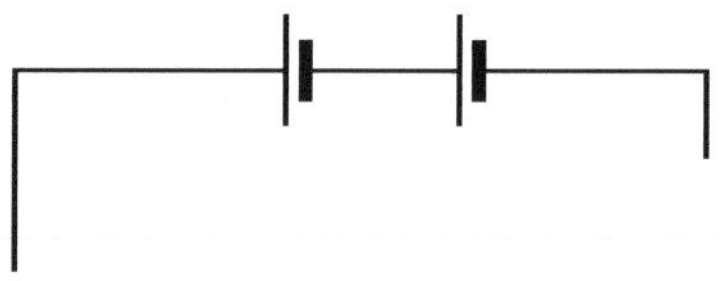

(3 marks)

4 Explain the shape of each graph shown below for different circuit components.

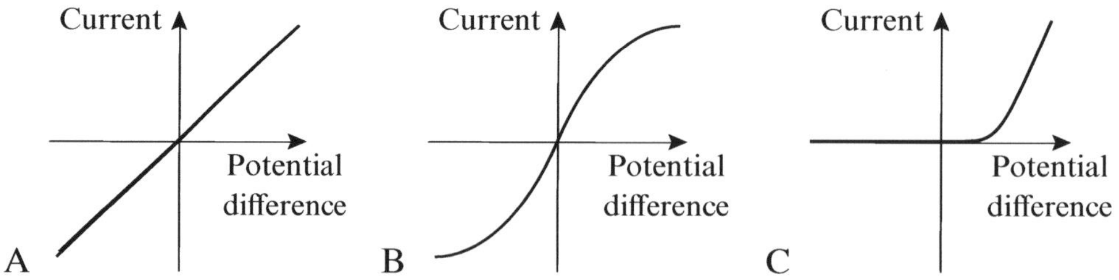

....

Guided

A Fixed resistor: The temperature remains .... so resistance remains

.... as shown by the straight line on the graph.   **(1 mark)**

B Filament lamp: As potential difference increases, the filament gets ....

so resistance .... as shown by the curved line on the graph.   **(1 mark)**

C Diode: The current flows in only .... and the resistance is ....

as shown by the curved line on the graph.   **(1 mark)**

Resistors

1   The graphs below (*I–V* graphs) show three types of component.

Current ▲   Current ▲

                        Potential   Potential
                        difference   difference

A   B

(a)   Describe what happens to the current through the component shown in **graph A** as the potential
      difference increases.

      ....

      ....

      .... **(2 marks)**

(b)   Describe what happens to the current through the component shown in **graph B** as the potential
      difference increases.

      ....

      ....

      .... **(2 marks)**

2   (a)   Complete the *I–V* graphs for a fixed resistor and a filament lamp.

         Fixed resistor   Filament lamp

         Current ▲   Current ▲

                        Potential   Potential
                        difference   difference

                                                                                        (2 marks)

   (b)   Explain why the filament lamp graph has a different shape to the fixed resistor graph
         (at constant temperature).

         ┌───┐
         │ A fixed resistor (at constant temperature) obeys Ohm's Law but a filament lamp does not. │
         └───┘

         ....

         ....

         .... **(2 marks)**

3   Describe an experiment to collect data to enable the calculation of the resistance of a wire.

    ▸ **Guided**   Data can be collected using an ammeter to measure ....

    and a voltmeter to measure ....

    A wire should be included and a fixed .... to prevent overheating. A range of

    .... measurements should be made so that resistance can be calculated

    using the equation .... **(5 marks)**

LDRs and thermistors

1 Draw the circuit symbols for the components in the boxes provided.

Light-dependent resistor (LDR)	Thermistor

(2 marks)

2 The sketch graphs below illustrate the relationship between two variables.

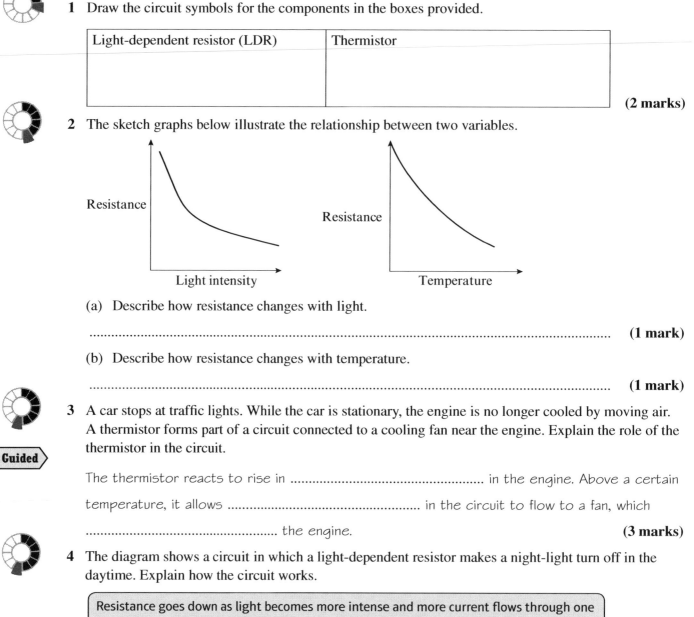

(a) Describe how resistance changes with light.

.... **(1 mark)**

(b) Describe how resistance changes with temperature.

.... **(1 mark)**

3 A car stops at traffic lights. While the car is stationary, the engine is no longer cooled by moving air. A thermistor forms part of a circuit connected to a cooling fan near the engine. Explain the role of the thermistor in the circuit.

> **Guided**

The thermistor reacts to rise in .... in the engine. Above a certain

temperature, it allows .... in the circuit to flow to a fan, which

.... the engine.   **(3 marks)**

4 The diagram shows a circuit in which a light-dependent resistor makes a night-light turn off in the daytime. Explain how the circuit works.

> Resistance goes down as light becomes more intense and more current flows through one component relative to the other, due to the way in which the circuit is wired.

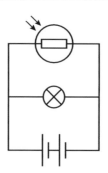

....

....

....

.... **(2 marks)**

16

Practical skills

Investigating *I–V* characteristics

1 (a) Complete the diagram below to show how the circuit could be used to investigate the *I–V* characteristics of a range of components.

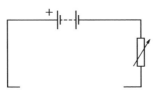

(2 marks)

(b) Suggest the type of data that should be collected using the circuit above by completing headings for the table below.

> These are the dependent and independent variables that you are recording in your experiment.

(i)	(ii)
Data	Data

(2 marks)

(c) Explain why the terminal connections should be reversed to collect additional data.

.... **(1 mark)**

(d) Write the labels you would give to the axes on your graph.

y-axis ....

x-axis .... **(1 mark)**

2 Suggest the component that has been tested by looking at the graphs of the data collected below.

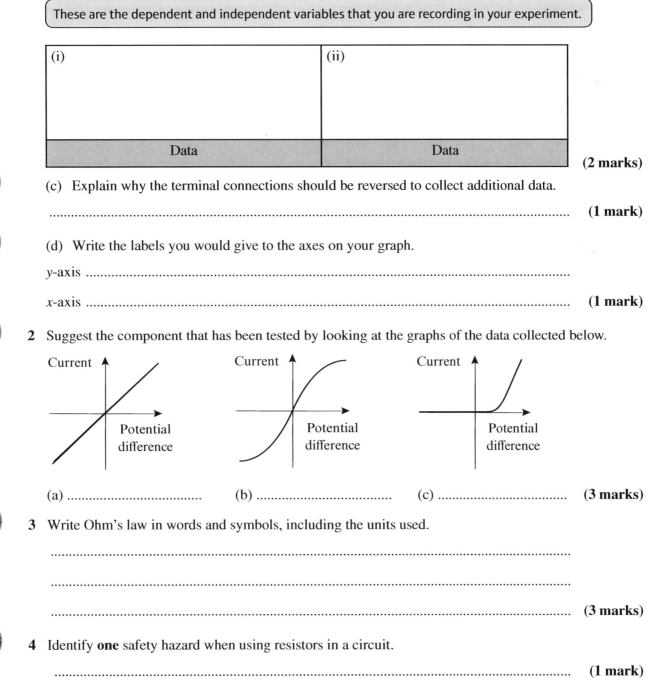

(a) ....   (b) ....   (c) .... **(3 marks)**

3 Write Ohm's law in words and symbols, including the units used.

....

....

.... **(3 marks)**

4 Identify **one** safety hazard when using resistors in a circuit.

.... **(1 mark)**

Series and parallel circuits

1 (a) Explain the rules for current in series and parallel circuits.

In a series circuit the current flowing through each component is ....

In a parallel circuit, the current is ....

the components.

(2 marks)

(b) Each lamp in these circuits is identical. Write the current for each ammeter on the circuit diagrams.

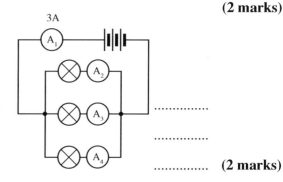

(2 marks)

2 (a) Explain the rules for potential difference in series and parallel circuits.

In a series circuit, the total potential difference supplied is .... the components.

In a parallel circuit, the potential difference across each component is ....

as the potential difference supplied.

(2 marks)

(b) Each lamp in these circuits is identical. Write the potential difference for each voltmeter, on the circuit diagrams.

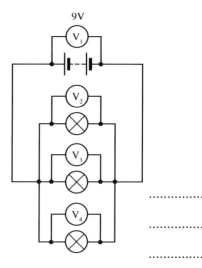

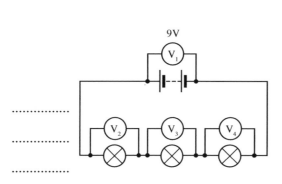

(2 marks)

3 Describe the difference between the total resistance of a series circuit and the total resistance of a parallel circuit, as illustrated below.

You may find the equation for series circuits $R = R_1 + R_2$ helpful in considering how current flows.

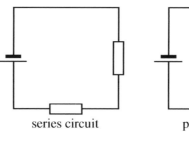

series circuit   parallel circuit

....

....

....

.... **(4 marks)**

ac and dc

1 Circuits can operate using either a direct potential difference and current or an alternating potential difference and current.

(a) Explain what is meant by **direct** potential difference and current.

Direct potential difference is constant and the current flows in the ....

direction.   **(2 marks)**

(b) Explain what is meant by **alternating** potential difference and current.

Alternating potential difference is .... and the current constantly

.... direction.   **(2 marks)**

2 Calculate the energy transferred for each of the following appliances:

Remember to convert units where appropriate.

(a) A fan heater (2000 W) running for 15 minutes.

....

....

Energy transferred = ... J **(2 marks)**

(b) A tablet charger (10 W) running for 6 hours.

....

....

Energy transferred = ... J **(2 marks)**

3 Complete the graph to show what the trace of a direct current would look like on an oscilloscope.

Volts

Time

(1 mark)

4 Calculate which appliance has the highest power rating:

(a) a toaster transferring 120 000 J in 60 seconds

....

....

or

(b) a kettle тransferring 252 000 J in 2 minutes.

....

.... **(5 marks)**

Mains electricity

1 Add labels to complete the diagram of a household plug.

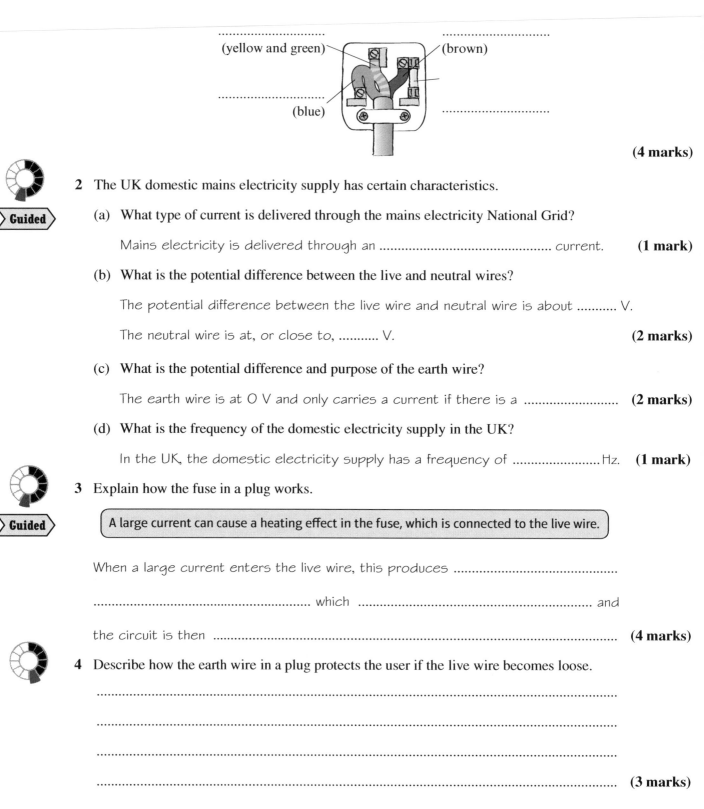

.... (yellow and green)

.... (brown)

.... (blue)

....

(4 marks)

2 The UK domestic mains electricity supply has certain characteristics.

> **Guided**

(a) What type of current is delivered through the mains electricity National Grid?

Mains electricity is delivered through an .... current.   **(1 mark)**

(b) What is the potential difference between the live and neutral wires?

The potential difference between the live wire and neutral wire is about .... V.

The neutral wire is at, or close to, ... V.   **(2 marks)**

(c) What is the potential difference and purpose of the earth wire?

The earth wire is at 0 V and only carries a current if there is a ....   **(2 marks)**

(d) What is the frequency of the domestic electricity supply in the UK?

In the UK, the domestic electricity supply has a frequency of ....Hz.   **(1 mark)**

3 Explain how the fuse in a plug works.

> **Guided**

> A large current can cause a heating effect in the fuse, which is connected to the live wire.

When a large current enters the live wire, this produces ....

.... which .... and

the circuit is then ....   **(4 marks)**

4 Describe how the earth wire in a plug protects the user if the live wire becomes loose.

....

....

....

....   **(3 marks)**

Electrical power

1 A hotplate is used to heat up a saucepan of water. The hotplate uses mains voltage of 230 V. The electric current through the hotplate is 5 A. Calculate the power of the hotplate in watts.

Guided

$P = IV = 5 \, A \times$ V

so $P =$ .... W   **(2 marks)**

2 The potential difference across a cell is 6 V. The cell delivers 3 W of power to a filament lamp.

(a) Calculate the current flowing through the lamp.

....

....

....

Current = ... A   **(3 marks)**

(b) A new lamp has a resistance of 240 Ω which draws a current of 0.5 A. Calculate the power rating of the new lamp.

> You need to recall the equation $P = I^2 R$

....

....

....

Power = ... W   **(3 marks)**

3 A coffee maker draws a current of 6 A using a mains voltage of 230 V. The coffee maker is switched on for 15 minutes. Which calculation could be used to determine the power of a coffee maker? Tick **one** box.

☐ Power = (6 × 230 × 15) / 15 (W)

☐ Power = 6 × 230 / 900 (W)

☐ Power = 6 × 230 / 15 (W)

☐ Power = 6 × 230 (W)   **(1 mark)**

4 An 80 V electric drill has a resistance of 8 Ω. Calculate the power of the drill.

....

....

....

Power = ... W   **(3 marks)**

Maths skills

Electrical energy

1 Calculate the amount of energy transferred to a 9-V lamp when a charge of 30 C is supplied.

....

....

Energy transferred = ... J   **(3 marks)**

2 A mobile phone has a battery that produces a potential difference of 4 V. When making a call, the phone uses a current of 0.2 A. A student makes a call lasting 30 seconds.

Guided

You need to recall the equations $Q = I\,t$ and $E = Q\,V$

(a) Calculate the charge flow during the phone call.

Charge flow, Q = ...

Charge flow = ...C   **(1 mark)**

(b) Calculate the energy transferred by the mobile phone while the call is made. Choose the correct unit from the box.

watts / W	joules / J	volts / V

Energy transferred, E = ...

....

Energy transferred = ... unit.................   **(3 marks)**

3 Explain how the energy of a circuit device is related to the following:

(a) Power:

....

....   **(1 mark)**

(b) Potential difference and current:

....

....

....

....   **(2 marks)**

4 (a) Calculate the charge flow of 600 J of electrical energy with a potential difference of 20 V and a current of 0.15 A.

....

Charge flow = ... C   **(2 marks)**

(b) Calculate how long it takes for a resistor to transfer the 600 J of electrical energy.

....

Time = ... s   **(2 marks)**

The National Grid

1 The National Grid transmits electricity from power stations at 400 000 volts (400 kV).

(a) Explain why this voltage is used to transmit electricity long distances.

> Remember that increasing the voltage decreases the current.

....

....

....

....

.... **(3 marks)**

(b) Explain how this system can be described as 'efficient'.

....

....

....

.... **(2 marks)**

(c) Give **one** hazard of transmitting electricity at 400 000 V.

.... **(1 mark)**

2 A power station generates an electric current of 20 000A at a voltage of 25 kV. Calculate the power generated in kilowatts.

....

....

Power generated = ... kW **(2 marks)**

3 Describe **two** ways in which energy losses from the National Grid may be reduced.

....

....

....

.... **(2 marks)**

4 Transformers are used at various places in the National Grid. Describe the role of transformers.

Guided

Step-up transformers are used to increase the ....

as it ....

Near homes, step-down transformers ....

to reduce ....

.... **(4 marks)**

Static electricity

Guided

1 A student gives a rubber balloon a negative charge by rubbing it on his jumper. He then holds it close to a wall. Explain why the balloon sticks to the wall.

balloon _ _ wall

The balloon has a negative charge.

The student causes a charge on the balloon by ....

from .... to the .... The negative charges on the

balloon repel the ....

...., inducing a

positive charge on the wall which attracts the negatively charged balloon.   **(4 marks)**

2 Which of the following particles can accumulate to give a static charge? Tick **one** box.

☐ Electrons on conductors   ☐ Protons on conductors

☐ Electrons on insulators   ☐ Protons on insulators   **(1 mark)**

3 Describe how a material becomes:

(a) positively charged

....   **(1 mark)**

(b) negatively charged.

....   **(1 mark)**

4 Explain why static charges build up on insulators but not on conductors.

Consider how particles move through the two types of material.

....

....

....

....

....

....

....

....

....   **(5 marks)**

Electric fields

1 Explain what is meant by the term electric field.

....

....

.... **(2 marks)**

2 Draw a diagram to show the electric field of a strong positive point charge.

(3 marks)

3 Describe how an electric field of a charged object changes with distance.

....

....

....

.... **(2 marks)**

4 Suggest in which direction (outwards or inwards) the following particles will move when placed in a point source electric field.

    (a) (i) A proton in a field generated by a positive point charge....................................... **(1 mark)**

         (ii) An alpha particle in a field generated by a positive point charge........................... **(1 mark)**

         (iii) A proton in a field generated by a negative point charge....................................... **(1 mark)**

    (b) Explain why these particles will accelerate in such a field.

> When a charged particle is in an electric field it will experience repulsion or attraction depending on its charge.

....

....

.... **(2 marks)**

5 A student says that 'an electrically charged insulator generates an electric field'. Explain whether the student is right or wrong.

Guided

The student is .... because the electrically charged insulator will

either be .... due to gaining or losing

.... The charged insulator then becomes a point source

creating an .... **(3 marks)**

Extended response – Electricity

Explain how a circuit can be used to investigate the change in resistance for a thermistor and a light-dependent resistor. Your answer should include a use for each component.

You will be more successful in extended response questions if you plan your answer before you start writing.

The question asks you to give a detailed explanation of how resistance changes in two types of variable resistor. Think about:

- How resistance in a resistor can be measured and calculated.
- The variable that causes a change in resistance in a thermistor.
- The variable that causes a change in resistance in a light-dependent resistor.
- The consequence to the circuit of a change in resistance in a component.
- Uses for thermistors and light-dependent resistors.

You should try to use the information given in the question.

....

....

....

....

....

....

....

....

....

....

....

....

....

....

....

....

.... **(6 marks)**

Density

1 The diagram below shows the three states of matter for a metal.

(a) Identify the three states of matter. Write the names in the boxes. **(1 mark)**

1.	2.	3.

(b) Describe how mass per unit volume varies with particle arrangement for **each** of the states of matter shown in the diagrams.

....

....

....

....

.... **(3 marks)**

(c) Describe how mass per unit volume relates to density, for **each** of the states of matter shown in the diagrams.

1. ....

....

2. ....

....

3. ....

.... **(3 marks)**

2 Which of these statements correctly describes density? Tick **one** box.

☐ Density is constant for a substance, whether it is a solid, a liquid or a gas.

☐ Density is calculated by dividing its mass by its volume.

☐ Density can be described by weight per unit area.

☐ Density is calculated by measuring force and volume. **(1 mark)**

3 A metal block measuring 10 cm × 25 cm × 15 cm has a density of 3 g/cm^3.

Maths skills

Calculate the mass of the block. Give your answer in kilograms.

> Work through the calculation first, then convert mass to kilograms at the end.

....

....

Mass of metal block = ... kg **(4 marks)**

Practical skills

Investigating density

1 When determining the density of a substance you need to measure the volume of the sample.

(a) Identify which other quantity you need to measure.

.... **(1 mark)**

(b) Give an example of how you could measure this quantity.

.... **(1 mark)**

2 (a) Describe a method that could be used to determine the volume of a regularly shaped solid.

....

....

....

.... **(2 marks)**

(b) Describe a method that could be used to determine the volume of an irregularly shaped solid.

....

....

....

.... **(2 marks)**

3 (a) Describe a method that could be used to find the density of a liquid using a balance.

Guided

Place a measuring cylinder on a balance and then zero the scales with ....

Add the liquid and .... Record the ....

(in g) from the balance and ....(in cm³) by reading from the level

in the measuring cylinder.   **(3 marks)**

(b) Describe the technique to read the volume of the liquid accurately.

....

....

.... **(2 marks)**

(c) Calculate the density of a liquid with a mass of 121 g and a volume of 205 cm³.

> You may find this equation useful
> $density = mass \div volume \quad (\rho = m / V)$

....

....

....

density = ... g/cm³ **(2 marks)**

Changes of state

1   Describe the difference between intermolecular forces between particles in a liquid and those in a gas.

....

....

.... **(2 marks)**

2   Describe **two** significant differences between the states of matter.

> Refer to the reasons why particles behave differently.

....

....

.... **(2 marks)**

3   Which statement describes the energy change when ice melts and then refreezes? Tick **one** box.

☐   Energy is transferred to surroundings → further energy is transferred to surroundings.

☐   Energy is transferred to the ice → energy is transferred to surroundings.

☐   Energy is transferred to surroundings → energy is transferred to the ice.

☐   Energy is transferred to the ice → energy remains in the system.   **(1 mark)**

4   Explain why the temperature stops rising when a liquid is heated to its boiling point and heating continues.

Guided

At boiling point, the ....

so the energy applied after boiling point is reached goes into ....

....

The particles .... and become a .... **(3 marks)**

5   When water is put into the freezer it turns to ice at 0 °C.

Explain what happens, in terms of the energy store, as the temperature continues to fall to −18 °C.

> Remember that the energy stores change as temperature falls.

....

....

....

....

....

.... **(3 marks)**

Internal energy

1   Choose the correct description of internal energy. Tick **one** box.

☐   The total number of particles in the system.

☐   The total sum of the kinetic and potential energies of the particles inside the system.

☐   The total number of energies inside the system.

☐   The total sum of the kinetic and potential energies of the particles outside the system.

(1 mark)

2   A liquid is heated at boiling point before becoming a gas. Describe the change, if any, of the potential and kinetic energies of the particles.

> Remember that latent heat of vaporisation occurs at the boiling point.

liquid   gas

....

....

....

....   **(2 marks)**

3   (a) Explain how internal energy changes when there is a temperature rise in a substance.

....

....

....

....   **(2 marks)**

   (b) Explain how internal energy changes when there is no change in temperature.

....

....

....

....   **(2 marks)**

4   Explain the condensation of water vapour in terms of kinetic and potential energies of the particles.

Guided

When the water vapour condenses into a liquid there will be no change in the ....

.... of the water particles so the temperature ....

.... but there will be a ....

in the .... of the water particles as they move from a gas

state to a liquid state.

(3 marks)

Specific latent heat

1 Identify what is meant by the term 'specific latent heat'.

.... **(1 mark)**

2 Which of the following remains constant during a change of state? Tick **one** box.

☐ Thermal energy of the mass

☐ Kinetic energy of the particles

☐ Temperature of the mass

☐ Energy supplied to the mass　　　　　　　　　　　　　　　　**(1 mark)**

3 Calculate the amount of energy needed to melt 25 kg ice. Take the specific latent heat of fusion of water to be 336 000 J/kg.

....

....

Energy required = ... J　**(2 marks)**

4 Which of the letters A–E on the graph correspond to the following stages during the temperature rise of water over time?

Temperature in °C

(a) Melting ....

(b) Boiling ....

(c) Specific latent heat of fusion ....

(d) Specific latent heat of vaporisation ....

(4 marks)

(e) Explain what is happening at stages B and D when there is no rise in temperature.

> Consider the bonds between particles.

....

.... **(2 marks)**

5 Calculate the energy needed to evaporate 36 kg of water at 100 °C to 36 kg steam at 100 °C.

Take the specific latent heat of vaporisation of water to be 2260 kJ/kg.

....

....

....

Energy needed = ... kJ　**(2 marks)**

Particle motion in gases

1 Suggest what is meant by temperature.

> Consider the movement of particles.

.... **(1 mark)**

2 Complete the table below showing some equivalent values in kelvin and degrees Celsius.   **(3 marks)**

kelvin (K)	degrees Celsius (°C)
	0
	−18
373	

3 The graph below represents the change in pressure with temperature of a gas at constant volume. Describe the relationship shown in the graph between pressure and temperature of the gas.

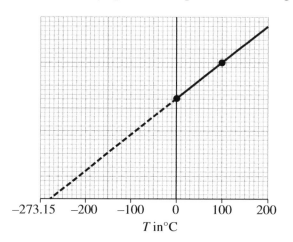

.... **(1 mark)**

4 In an experiment, a fixed-volume container of 100 g of helium gas is warmed from −10 °C to 30 °C.

   (a) Describe what happens to the velocity of the helium particles as a result of increasing temperature.

Guided

As the temperature increases the particles will move ....

because they gain more .... **(2 marks)**

   (b) Explain how this affects the pressure on the container walls.

....

....

....

.... **(2 marks)**

   (c) State what happens to the average kinetic energy of the particles as the temperature increases.

.... **(1 mark)**

Pressure in gases

1 Describe the pressure of a gas on a surface in terms of particle movement and force.

When particles of a gas collide with a surface they ....

....

resulting in .... **(3 marks)**

2 When a gas is at constant temperature, what is the relationship between volume and pressure?
Tick **one** box.

☐ Pressure is directly proportional to volume.

☐ Volume and pressure both decrease.

☐ Volume is inversely proportional to pressure.

☐ Pressure and volume both increase.   **(1 mark)**

3 The pressure in a cylinder of air is increased from atmospheric pressure (100 kPa) to 280 kPa.

The original volume of the cylinder is 230 cm^3.

Calculate the volume of the cylinder when the pressure is 280 kPa.

> Rearrange the equation $P_1 V_1 = P_2 V_2$

....

....

....

....

Volume = ... cm^3 **(3 marks)**

4 A mixture of nitrous oxide and oxygen gases is used as an anaesthetic by dentists; 640 litres of the gas at normal atmospheric pressure (100 kPa) can be stored in a cylinder with an internal volume of 8 litres.

> Remember to allow for the 8-litre tank – pressure will be equalised only when 8 litres of gas remains.

Calculate the pressure needed to compress the gas into the cylinder.

....

....

....

....

Pressure = ... Pa   **(3 marks)**

Extended response – Particle model

Substances can undergo a change of state. Explain, using the kinetic particle theory, the changes of state of water. In your answer, include reasons that explain why thermal energy input and output are not always linked to changes in temperature.

> You will be more successful in extended response questions if you plan your answer before you start writing.
>
> The question asks you to give a detailed explanation of how water changes state in terms of particles. Think about:
>
> - The relative kinetic energy of particles in solids, liquids and gases.
> - Why a change of state is described as a reversible change.
> - How heating the system can result in a change in temperature.
> - Why heating the system does not always result in a change in temperature.
> - How latent heat is involved in the process.
> - Include equations to help explain your answer.
>
> You should try to use the information given in the question.

....

....

....

....

....

....

....

....

....

....

....

....

....

....

....

....

....

....

.... **(6 marks)**

The structure of the atom

1 Complete the diagram to show the location and charge of:

   (a) protons   **(1 mark)**

   (b) neutrons   **(1 mark)**

   (c) electrons   **(1 mark)**

2 (a) Explain why atoms have no overall charge.

    The number of .... in the nucleus is

    .... orbiting the nucleus. **(2 marks)**

   (b) Describe what will happen to the overall charge if an atom loses an electron.

    .... **(1 mark)**

3 Identify the approximate diameter of an atom and a nucleus from the measurements in the box.

10^{-18} m	10^{-15} m	10^{-10} m	10^{-9} m	10^{-6} m	10^{-2} m

    Size of an atom: ....

    Size of a nucleus: ....   **(2 marks)**

4 The diagram below shows the emission of electromagnetic radiation from an atom. Describe the process of absorption and emission that would cause this to occur.

electron

> The diagram shows emission but you will also need to explain the absorption that happens first.

    ....

    ....

    ....

    ....

    ....

    ....

    ....

    .... **(4 marks)**

Atoms, isotopes and ions

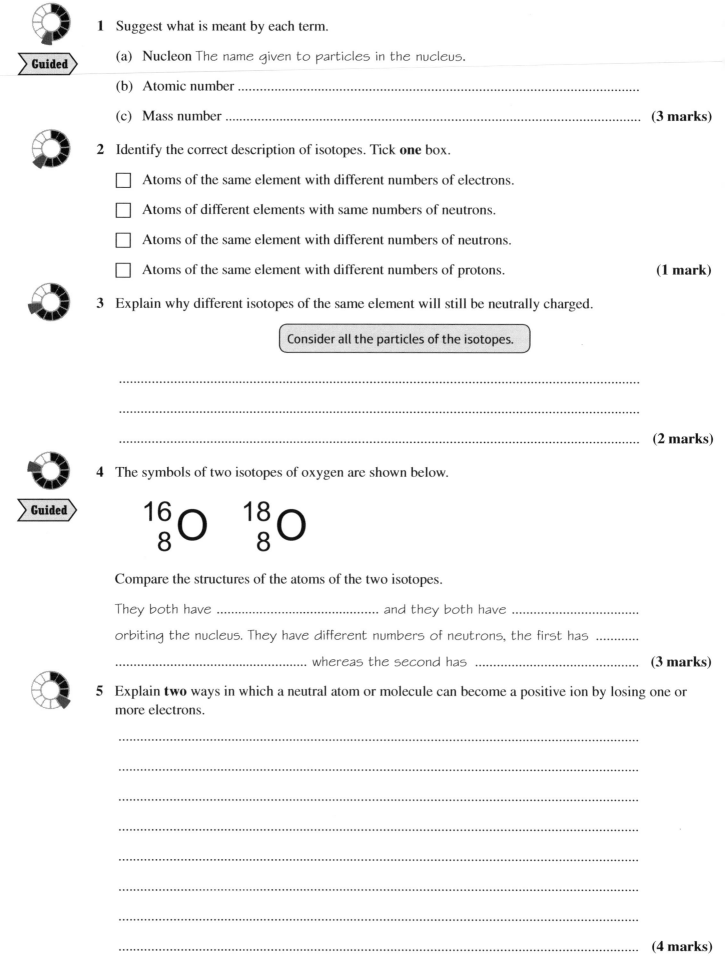

1 Suggest what is meant by each term.

(a) **Nucleon** The name given to particles in the nucleus.

(b) Atomic number ....

(c) Mass number .... **(3 marks)**

2 Identify the correct description of isotopes. Tick **one** box.

☐ Atoms of the same element with different numbers of electrons.

☐ Atoms of different elements with same numbers of neutrons.

☐ Atoms of the same element with different numbers of neutrons.

☐ Atoms of the same element with different numbers of protons. **(1 mark)**

3 Explain why different isotopes of the same element will still be neutrally charged.

> Consider all the particles of the isotopes.

....

....

.... **(2 marks)**

4 The symbols of two isotopes of oxygen are shown below.

$$^{16}_{8}\text{O} \qquad ^{18}_{8}\text{O}$$

Compare the structures of the atoms of the two isotopes.

They both have .... and they both have ....

orbiting the nucleus. They have different numbers of neutrons, the first has ....

.... whereas the second has .... **(3 marks)**

5 Explain **two** ways in which a neutral atom or molecule can become a positive ion by losing one or more electrons.

....

....

....

....

....

....

.... **(4 marks)**

Models of the atom

1 Compare the plum pudding and Rutherford models of the atom.

The plum pudding model showed the atom as ....

particle containing ....

whereas the Rutherford model showed the atom as ....

.... surrounded by .... **(4 marks)**

2 Describe the evidence that enabled Rutherford to make his claim about the nucleus.

....

....

....

....

....

.... **(3 marks)**

3 (a) Bohr developed the Rutherford model by describing orbits around the nucleus. Which particle did Bohr propose that travelled in this way? Tick **one** box.

☐ The electron

☐ The proton

☐ The neutron

☐ The positron   **(1 mark)**

(b) Explain how the Bohr model of the atom improved on Rutherford's model.

....

....

....

....

....

....

....

.... **(4 marks)**

Radioactive decay

1 Complete the following sentences to explain the radioactive process.

(a) Activity is the .... nuclei decay per second. **(2 marks)**

(b) The unit of activity is the .... **(1 mark)**

(c) Count rate is the .... **(2 marks)**

2 Identify the equipment used to detect the activity of a radioactive source. Tick **one** box.

☐ MRI scanner

☐ PET scanner

☐ X-ray generator

☐ Geiger–Müller (GM) tube   **(1 mark)**

3 A radioactive material has an activity of 450 Bq. Calculate how many radioactive nuclei will decay in 2 minutes.

Guided

In 1 second, ... will decay,

so .... = ... nuclei will decay in 2 minutes. **(3 marks)**

4 In radioactive decay, changes occur in the nucleus. The changes resulting from two different types of radioactive decay are described below. Describe the nature of the radiation, including its name.

> The types of radiation to think about are:
> α, β–, β+, γ and neutron radiation.

(a) The nucleons are reduced by 4.

....

....

....

.... **(2 marks)**

(b) The positive charge of the nucleus increases by 1.

....

....

....

.... **(2 marks)**

Nuclear radiation

1 Select the correct description of an alpha particle. Tick **one** box.

☐ Helium nucleus with charge −2   ☐ High-energy neutron

☐ Helium nucleus with charge +2   ☐ Ionising electron   **(1 mark)**

2 Complete the table below.

Type of radiation	Penetrating power
	very low, stopped by 10 cm of air
	low, stopped by thin aluminium
	very high, stopped by very thick lead

(2 marks)

3 An atom of carbon-14, with 6 protons and 8 neutrons, undergoes beta-minus decay.

It becomes an atom of nitrogen, with 7 protons and 7 neutrons.

(a) Give the change in relative atomic mass.

.... **(1 mark)**

(b) Give another description for a beta-minus particle.

.... **(1 mark)**

(c) Name the relative ionising category of a beta-minus particle.

.... **(1 mark)**

4 Explain why alpha particles have the shortest ionising range in air, compared with other types of ionising radiation.

Compared with other types of ionising radiation, the chance of collision with air particles

....

....

because ....

.... **(3 marks)**

5 Explain why the use of dense shielding materials could reduce ionisation in the body by radioactive particles and waves.

> Ionisation occurs when electrons are knocked away from atoms, leaving a charged ion.

....

....

....

....

....

....

....

.... **(4 marks)**

Uses of nuclear radiation

1 Machinery that produces standard sheets of paper uses radiation to check the thickness.

> Recall the properties of alpha, beta and gamma radiation.

(a) Explain which type of radiation is used.

....

.... **(3 marks)**

(b) A radiation detector measures a sudden drop in the radiation that passes through the paper.

(i)   Give the most likely cause of this change.

.... **(1 mark)**

(ii) Suggest how the machine would respond to the change.

.... **(1 mark)**

2 Gamma-rays are used to treat cancers. Describe the property of gamma-rays that makes them useful for treating cancers.

.... **(1 mark)**

3 The diagram of the smoke alarm shows that the radioactive isotope americium-241 is used in the system. It is a source of alpha radiation.

> Guided

(a) Explain why it is safe to use smoke detectors in the home.

Alpha particles cannot pass through ....

and they are ....

.... **(2 marks)**

(b) Explain why the siren sounds when smoke gets into the smoke alarm.

The smoke particles absorb the alpha

particles ....

....

....

....

....

....

....

....

smoke entering
the alarm   americium-241

                                                         siren

air molecules

charged plate   detector

                                            battery

(2 marks)

4 Explain why some surgical instruments are irradiated with gamma-rays.

....

....

....

.... **(2 marks)**

Nuclear equations

1 Give the symbols used in decay equations to represent the following particles.

(a) Alpha particle: .... **(1 mark)**

(b) Beta-minus particle: .... **(1 mark)**

2 Radium-222 undergoes alpha decay. Identify which of the following statements is true. Tick **one** box.

☐ The positive charge of the nucleus is reduced by 4.

☐ The atomic number is increased by 1.

☐ The nucleus gains an extra proton.

☐ The mass number is reduced by 4. **(1 mark)**

3 Using the data in the table, complete the equations below.

7	9	11	12	14	16	19	20
Li	Be	B	C	N	O	F	Ne
lithium	beryllium	boron	carbon	nitrogen	oxygen	fluorine	neon
3	4	5	6	7	8	9	10
23	24	27	28	29	31	35.5	40
Na	Mg	Al	Si	P	S	Cl	Ar
sodium	magnesium	aluminium	silicon	phosphorus	sulfur	chlorine	argon
11	12	13	14	15	15	17	18

(a) $^{14}_{6}\text{C} \rightarrow ^{14}_{....}\text{N} + ^{0}_{-1}\text{e}$ **(1 mark)**

(b) $^{23}_{....}\text{Mg} \rightarrow ^{23}_{11}\text{Na} + ^{0}_{1}\text{e}$ **(1 mark)**

4 Identify the type of radiation that would be emitted in each decay.

> Remember the law of conservation of mass.

(a) Carbon-10 (6 protons, 4 neutrons) → boron-10 (5 protons, 5 neutrons)

.... **(1 mark)**

(b) Uranium-238 (92 protons, 146 neutrons) → thorium-234 (90 protons, 144 neutrons)

.... **(1 mark)**

(c) Helium-5 (2 protons, 3 neutrons) → helium-4 (2 protons, 2 neutrons)

.... **(1 mark)**

5 Complete each equation, naming the type of decay shown.

> Check that the A and Z numbers obey the conservation laws.

(a) $^{....}_{84}\text{Po} \rightarrow ^{4}_{2}\text{He} + ^{204}_{82}\text{Pb}$ Type of decay .... **(2 marks)**

(b) $^{222}_{....}\text{Rn} \rightarrow ^{4}_{2}\text{He} + ^{218}_{84}\text{Po}$ Type of decay .... **(2 marks)**

(c) $^{42}_{19}\text{K} \rightarrow ^{0}_{-1}\text{e} + ^{....}_{20}\text{Ca}$ Type of decay .... **(2 marks)**

(d) $^{....}_{4}\text{Be} \rightarrow ^{1}_{0}\text{n} + ^{8}_{4}\text{Be}$ Type of decay .... **(2 marks)**

Half-life

1 Define the term 'half-life'.

....

.... **(1 mark)**

2 A sample of thallium-208 contains 16 million atoms. Thallium-208 has a half-life of 3.1 minutes.

(a) Give the number of nuclei that will have decayed in 3.1 minutes.

Number of atoms = ... **(1 mark)**

(b) Calculate the number of unstable thallium nuclei left after 9.3 minutes.

> Remember that thallium-208 has a half-life of 3.1 minutes, so there are three half-lives in 9.3 minutes.

Maths skills

....

....

Number of thallium nuclei after 9.3 min = ... **(2 marks)**

3 A student measured the activity of a radioactive sample for 30 minutes. She plotted the graph of activity against time shown on the right.

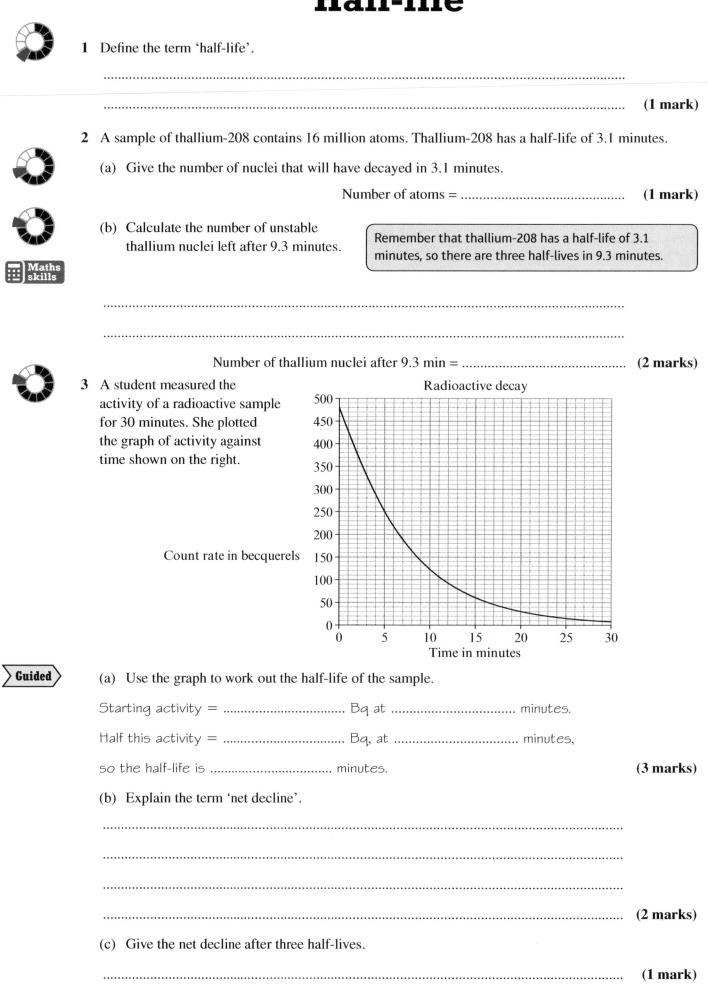

Radioactive decay

Count rate in becquerels

Time in minutes

Guided

(a) Use the graph to work out the half-life of the sample.

Starting activity = ... Bq at .... minutes.

Half this activity = ... Bq, at .... minutes,

so the half-life is .... minutes.   **(3 marks)**

(b) Explain the term 'net decline'.

....

....

....

.... **(2 marks)**

(c) Give the net decline after three half-lives.

.... **(1 mark)**

Contamination and irradiation

1 During the First World War (1914–18) soldiers and airmen were issued with watches that had hands and numbers that glowed in the dark. The hands and numbers had been painted with luminous paint that contained radium. Radium was discovered in 1898 and found to be radioactive. In the 1920s, many of the women who painted the watches became very ill. Discuss why it was not banned from being used on watches until the 1920s.

Guided

Before 1920, the effects of radium ....

so it was thought ....

It was banned from use .... **(3 marks)**

2 Define the following terms about exposure to radiation.

(a) External contamination

....

.... **(1 mark)**

(b) Internal contamination

....

.... **(1 mark)**

(c) Irradiation

....

.... **(1 mark)**

3 Give an example of how a person may be subjected to:

(a) external contamination

.... **(1 mark)**

(b) internal contamination

.... **(1 mark)**

4 Explain why alpha particles are more dangerous from a source of internal contamination than external contamination.

> Alpha particles travel only a very short distance before colliding with another particle, losing their energy. This can have serious consequences near to the body.

....

....

....

....

.... **(4 marks)**

Hazards of radiation

1   Identify the most ionising radiation. Tick **one** box.

☐ Alpha rays

☐ Beta rays

☐ Gamma-rays

☐ X-rays   **(1 mark)**

2   Suggest **three** precautions that may be taken by people who may come into contact with ionising radiation to reduce the risk of cell damage.

1. ....

2. ....

3. .... **(3 marks)**

3   Explain why alpha particles are more dangerous inside the body than gamma-rays.

Guided

A source of alpha particles with high activity inside the body ....

because they .... before transferring all of their ionising

energy. Gamma-rays can .... without ....

.... **(4 marks)**

4   When radioactive sources are handled, tongs are often used. Explain why those handling radioactive sources would use tongs.

radioactive source

....

.... **(2 marks)**

5   X-rays can also be ionising to body cells. Suggest why it is considered safe for the patient to be exposed to the X-rays whereas the medical workers have to leave the room.

....

....

.... **(3 marks)**

Background radiation

1 The pie chart shows the sources of background radiation, 50% of which comes from the element radon.

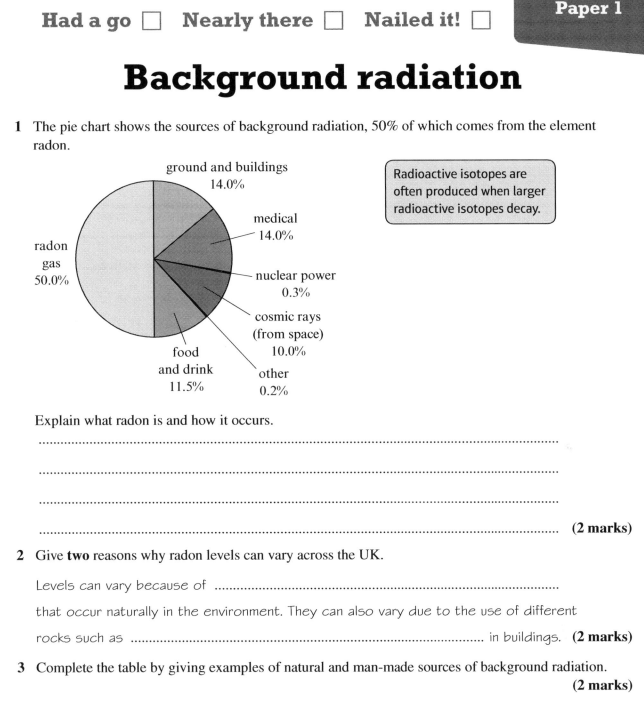

ground and buildings
14.0%

medical
14.0%

nuclear power
0.3%

cosmic rays
(from space)
10.0%

other
0.2%

food
and drink
11.5%

radon
gas
50.0%

Radioactive isotopes are often produced when larger radioactive isotopes decay.

Explain what radon is and how it occurs.

....

....

....

.... **(2 marks)**

2 Give **two** reasons why radon levels can vary across the UK.

Guided

Levels can vary because of ....

that occur naturally in the environment. They can also vary due to the use of different

rocks such as .... in buildings. **(2 marks)**

3 Complete the table by giving examples of natural and man-made sources of background radiation.

(2 marks)

Sources of background radiation	
Natural	**Man-made**

4 Two scientists in different parts of the country measure the background radiation count three times. Their results are shown in the table below.   **(2 marks)**

Test number	1	2	3	Average
South East activity (Bq)	0.30	0.24	0.27	
South West activity (Bq)	0.31	0.28	0.32	

(a) Calculate the average activity for each sample and write it in the table.

.... **(1 mark)**

(b) Suggest which area has the highest average level of background radiation.

.... **(1 mark)**

Medical uses

1 Gamma radiation can be used in medical centres to destroy unwanted tissue. Which of these statements explains why gamma radiation is used? Tick **one** box.

☐ Gamma radiation is more easily absorbed by cells than alpha or beta radiation.

☐ Gamma radiation penetrates deeper from outside the body than alpha or beta radiation.

☐ Gamma radiation is applied using a wide beam of radiation.

☐ Radioactive isotopes used have a very long half-life.   **(1 mark)**

2 Give **three** of the safety precautions taken by people who work with ionising radiation each day.

1. ....

....

2. ....

....

3. ....

.... **(3 marks)**

3 Explain what is meant by a medical tracer, including how it is used.

> **Guided**

A medical tracer is a ....

It is injected into the patient and is then ....

....

A special camera detects the ....

The detected waves are used to build up an image of ....

.... **(4 marks)**

4 The radioactive isotope used in medicine to produce gamma radiation must be carefully selected to have the appropriate half-life. Suggest **two** reasons why half-life is important in medicine.

....

....

....

.... **(2 marks)**

Nuclear fission

1 Label the particles shown in the diagram of nuclear fission. Use the words in the box below.

| neutrons | daughter nuclei | uranium-235 |

energy release

(3 marks)

2 (a) Define the term 'nuclear fission'.

....

.... **(2 marks)**

(b) Spontaneous fission is rare. Explain what must usually happen for fission to occur.

....

.... **(2 marks)**

3 (a) Describe what is meant by the term 'chain reaction'.

....

....

....

.... **(3 marks)**

(b) Explain the measures taken to control a chain reaction in a nuclear reactor.

> Control measures are used to stop the process going out of control.

....

....

.... **(2 marks)**

(c) Draw a diagram to illustrate a chain reaction.

(4 marks)

(d) Identify where controlled chain reactions are used.

.... **(1 mark)**

Nuclear fusion

1 Which statement is a correct example of fusion? Tick **one** box.

☐ Helium nuclei → hydrogen nuclei

☐ Hydrogen nuclei → helium nuclei

☐ Uranium nuclei → thorium nuclei

☐ Plutonium → uranium nuclei   **(1 mark)**

2 In 1989 Martin Fleischmann and Stanley Pons announced that they had performed nuclear fusion in a process that newspaper reporters called 'cold fusion'. Since then, however, other scientists have been unable to repeat the experiment.

Explain why it has been difficult for Fleischmann and Pons to prove that their experiments were successful.

> Recall the scientific method and how it is applied.

....

....

....

.... **(2 marks)**

3 This equation shows the fusion of deuterium and tritium that occurs in stars:

$$^2_1H + {}^3_1H \rightarrow {}^4_2He + {}^1_0n + energy$$

Explain why energy is given out during this reaction.

> **Guided**

When fusion occurs the mass of the products is ....

so .... **(2 marks)**

4 (a) Explain why the process of nuclear fusion is so difficult to achieve.

....

....

.... **(2 marks)**

(b) Explain what would be needed to overcome the difficulties of the nuclear fusion process.

....

....

....

....

.... **(3 marks)**

Extended response – Radioactivity

Ionising radiation will travel through some materials but will be stopped by others. The diagram shows three different materials and how alpha, beta or gamma radiation may be absorbed by each one.

Use the diagram to help you explain about the structure and properties of these three types of radiation. In your answer, name the materials that could be shown in the diagram, and give other examples of how these types of radioactivity may be absorbed.

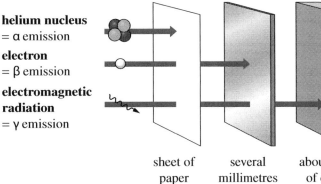

helium nucleus = α emission

electron = β emission

electromagnetic radiation = γ emission

sheet of paper   several millimetres of aluminium   about 1 metre of concrete or lead

> You will be more successful in extended response questions if you plan your answer before you start writing.
>
> The question asks you to give a detailed explanation of the penetrative characteristics of ionising radiation and examples of materials that may stop the radiation. Think about:
>
> - The relative ionising abilities of alpha, beta and gamma radiation.
> - How energy is transferred when radiation encounters a particle.
> - The effect that particle collisions have on how radiation passes through a material.
> - Examples of materials that absorb different types of radiation.
>
> You should try to use the information given in the diagram.

....

....

....

....

....

....

....

....

....

....

....

.... **(6 marks)**

Scalars and vectors

1 (a) Complete the table using the examples of scalars and vectors from the box below.

| acceleration | displacement | speed | energy | temperature |
| mass | force | velocity | momentum | distance |

Scalars	Vectors

(2 marks)

(b) Give one example of a scalar from the table, and explain why it is a scalar.

.... is a scalar because it has a size / magnitude but no ....

....

.... (3 marks)

2 Two students set off jogging in opposite directions. The first student starts to jog to the east at a velocity of 2 m/s. The second student jogs to the west at a velocity of −2 m/s.

N

W   E

2 m/s   ←   2 m/s   →

S

(a) (i) Explain why velocity is used in this example rather than speed.

.... (1 mark)

(ii) Explain why the velocity for the student jogging to the west is given a negative value.

.... (1 mark)

(b) Explain the importance of the length of the diagram arrows. Describe how you would draw the arrow if student 2 sped up to 3 m/s.

....

....

....

.... (2 marks)

3 (a) Which of the following is **not** a scalar? Tick **one** box.

☐ Energy   ☐ Temperature   ☐ Mass   ☐ Weight

(1 mark)

(b) Give a reason for your answer to (a).

.... (1 mark)

Interacting forces

1 Name the three types of fields that cause objects to interact with each other without making contact.

.... **(3 marks)**

2 Which of the following statements correctly describes the similarities between magnetic and electrostatic fields? Tick **one** box.

☐ Like poles and charges repel.

☐ Like poles and charges attract.

☐ Opposite poles and charges have a null point.

☐ Opposite poles and charges repel.   **(1 mark)**

3 Explain why weight and normal contact force are described as vectors and how they differ from each other.

Guided

Weight and normal contact forces are vectors because ....

....

Weight is measured .... whereas normal contact force is measured

.... **(3 marks)**

4 A student pulls along a luggage bag, as shown in the diagram, at constant velocity.

(a) Name the contact forces that are balanced for the horizontal plane.

.... **(1 mark)**

(b) Name the balanced contact forces in the vertical direction.

.... **(1 mark)**

5 A skydiver jumps from an aeroplane. Explain which forces influence the descent of the skydiver and how the net result of the forces controls how the skydiver descends, at each stage.

Make sure your answer includes the terms air resistance, terminal velocity and speed.

....

....

....

....

....

.... **(4 marks)**

Gravity, weight and mass

1 The lunar roving vehicle (LRV) has a mass of 210 kg on Earth.

(a) Give the mass of the unchanged LRV on the Moon and a reason for your answer.

Guided

The mass of the LRV on the Moon is .... kg because...........................

.... **(2 marks)**

(b) Draw an arrow on the LRV to show where the centre of mass would act and explain what is meant by centre of mass (in the space below).

.... **(2 marks)**

2 Calculate the total weight of a backpack of mass 1 kg, containing books with a mass of 2 kg and trainers with a mass of 1.5 kg. Take gravitational field strength (*g*) to be 10 N/kg.

Maths skills

> You will find the equation $W = m\,g$ useful.

....

....

....

....

Weight = ... N   **(2 marks)**

3 Kate is about to fly to Europe. The total baggage allowance is 20 kg. Kate has scales that weigh only in newtons. Determine the items that Kate can take on holiday, as well as her clothes, to get as close to 20 kg as possible. Show your calculations. Take gravitational field strength to be 10 N/kg.

> Remember to convert the weights shown to masses before you add them up.

laptop 45 N	camera bag 55 N	walking boots 25 N	jacket 35 N	clothes 105 N

....

....

....

....

Total baggage = ... kg   **(3 marks)**

Resultant forces

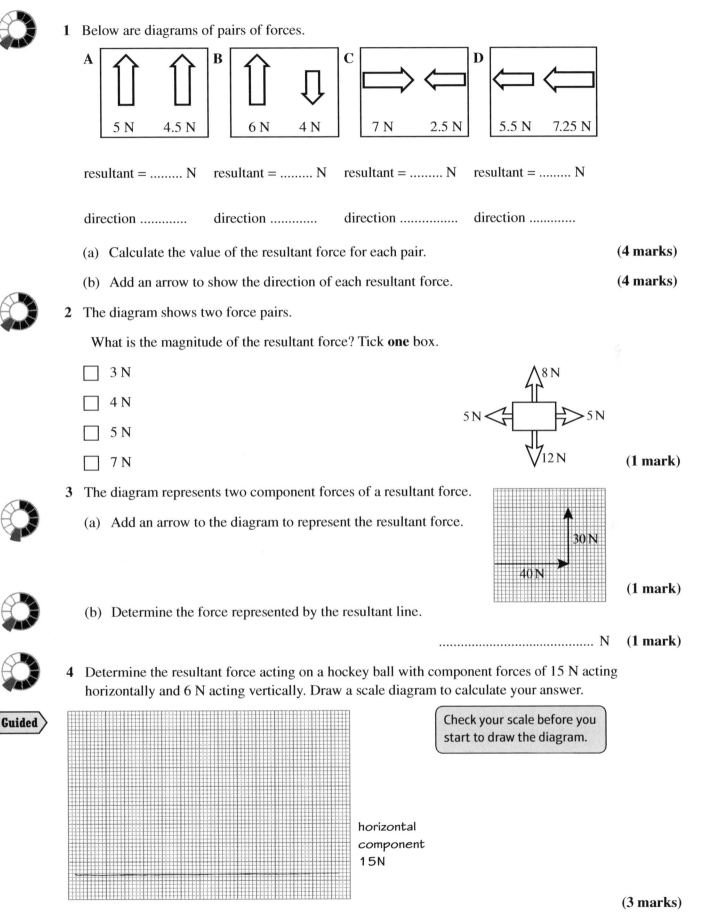

1 Below are diagrams of pairs of forces.

A | 5 N   4.5 N   B | 6 N   4 N   C | 7 N   2.5 N   D | 5.5 N   7.25 N

resultant = ... N   resultant = ... N   resultant = ... N   resultant = ... N

direction ....   direction ....   direction ....   direction ....

   (a) Calculate the value of the resultant force for each pair.   **(4 marks)**

   (b) Add an arrow to show the direction of each resultant force.   **(4 marks)**

2 The diagram shows two force pairs.

   What is the magnitude of the resultant force? Tick **one** box.

   ☐   3 N

   ☐   4 N

   ☐   5 N

   ☐   7 N   **(1 mark)**

3 The diagram represents two component forces of a resultant force.

   (a) Add an arrow to the diagram to represent the resultant force.

                                                       (1 mark)

   (b) Determine the force represented by the resultant line.

                                       .... N   **(1 mark)**

4 Determine the resultant force acting on a hockey ball with component forces of 15 N acting horizontally and 6 N acting vertically. Draw a scale diagram to calculate your answer.

Guided

> Check your scale before you start to draw the diagram.

horizontal component 15N

                                                 (3 marks)

Free-body force diagrams

 1 Which of the following is important for resolving a resultant force using a scale drawing? Tick **one** box.

☐ The magnitude of forces are not relevant for drawing lines.

☐ Each line on the drawing must be measured carefully.

☐ The vertical component can be any length.

☐ The horizontal line is not necessarily drawn to scale. **(1 mark)**

2 The diagram shows a bird of prey of mass 2 kg on a branch of a tree.

> You will need to recall the equation $F = m\,a$, where a can be rounded to 10 m/s².

(a) Add arrows to the diagram to show the direction and relative magnitude of forces acting on the bird.

(2 marks)

(b) Add the magnitude and units for each force on the diagram. **(2 marks)**

 3 The free-body diagram below represents a cyclist. Add arrows to represent the cyclist accelerating to the left along a horizontal path.

> **Guided**

> Remember to consider the relative length of the arrows drawn.

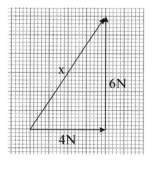

(2 marks)

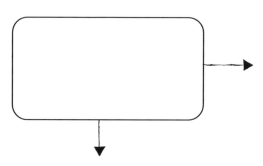 **4** The diagram shows the horizontal and vertical components of a force.

(a) Measure the length of the line of the resultant force.

.... cm **(1 mark)**

(b) Give the magnitude of the resultant force.

.... N **(1 mark)**

Work and energy

1 Which of these is the energy equivalent to the work done when a force is moved through a vertical distance? Tick **one** box.

☐ Gravitational potential energy ☐ Latent heat
☐ Kinetic energy ☐ Specific heat capacity **(1 mark)**

2 (a) Explain why friction may result in wasted energy when moving an object through a distance.

Work done against friction will lead to ....

....

which is dissipated to ....

.... **(2 marks)**

(b) Explain the consequence of friction in terms of work needed to be done in moving an object.

The greater the amount of friction ....

....

to move the body through ....

.... **(2 marks)**

3 A suitcase is pulled along a walkway with a force of 80 N. The work done in moving the case is 4800 J. Calculate the distance that the suitcase is moved.

> You should be able to recall the equation $W = F\,s$; it won't be provided on the equation sheet.

....

....

Distance = ... m **(2 marks)**

4 Cassie designs a pulley system to lift a mass of 8000 g.

The energy transferred by the force of the pulley is 320 J.

(a) Calculate the height to which the pulley system lifts the mass. Take g to be 10 N/kg.

> You should be able to recall the equation $E = m\,g\,h$; it won't be provided on the equation sheet.

....

....

Height = ... m **(2 marks)**

(b) Use your answer from (a) to calculate the force exerted by the pulley system.

....

....

Force = ... N **(3 marks)**

Forces and elasticity

1 Identify the force that would act opposite to the forward motion of a bicycle. Tick **one** box.

☐ Upthrust ☐ Friction

☐ Reaction force ☐ Tension **(1 mark)**

2 Give an example where each of the following may occur:

(a) Tension

.... **(1 mark)**

(b) Compression

.... **(1 mark)**

(c) Elastic distortion

.... **(1 mark)**

(d) Inelastic distortion

.... **(1 mark)**

3 A student investigates loading two aluminium beams, each with an elastic limit at 50 N. Beam 1 is tested to 45 N. Beam 2 is tested to 60 N. Predict what you would expect the beams to look like after loading.

> **Guided**

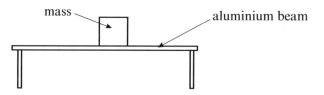

mass aluminium beam

Explain your answer.

After testing, Beam 1 would return to the same size and shape as ....

....

and would be intact. Beam 2 would ....

but would (probably) .... **(4 marks)**

4 Car manufacturers use inelastic distortion to make cars safer.

Discuss the function of the features that use inelastic distortion for safety reasons.

> Think of how energy is absorbed to protect the passengers in the event of a crash.

....

....

....

....

.... **(3 marks)**

Force and extension

1 Describe the difference between elastic deformation and inelastic deformation caused by stretching forces.

....

....

....

....

....

....

....

.... **(4 marks)**

2 A spring is stretched from 0.03 m to 0.07 m, within its elastic limit. Calculate the force needed to stretch the spring. Select the correct unit from the box below. Take the spring constant to be 80 N/m.

kg	m	N

Guided

Extension = 0.07 m - .... m = ... m

Maths skills

Force = ... × extension = ... × ....

Force = ... unit .... **(3 marks)**

3 Identify the spring constant that produces an extension of 0.04 m when a mass of 2 kg is suspended from a spring. Tick **one** box. (Take g to be 10 N/kg.)

☐ 0.02 N/m   ☐ 50 N/m

☐ 0.08 N/m   ☐ 500 N/m

> Be careful not to confuse mass with force.

(1 mark)

4 (a) Calculate the spring constant (k) of a spring that is stretched 15 cm when a force of 30 N is applied.

Maths skills

....

....

k = ... N/m **(3 marks)**

   (b) Calculate the energy transferred to the spring in (a).

> You need to select the equation:
> energy = ½ × spring constant × extension²
> from the Physics Equations Sheet.

....

....

Energy transferred = ... J **(2 marks)**

Forces and springs

1 (a) Describe how to set up an experiment to investigate the elastic potential energy stored in a spring using a spring, a ruler, masses or weights, a clamp and a stand.

> Include a step to make sure the spring is not damaged during the experiment.

....

....

....

....

....

.... **(4 marks)**

(b) Explain why it is important to check that the spring is not damaged during the experiment.

....

....

.... **(2 marks)**

(c) Explain how the data collected must be processed before a graph can be plotted. Assume masses are used and measurements made in mm.

> **Guided**

Masses must be converted to ....

The extension of the spring must be ....

....

Extension measurements should be .... **(3 marks)**

(d) Describe how a graph plotted from this experiment can be used to calculate:

  (i) the elastic potential energy stored in the spring

  .... **(1 mark)**

  (ii) the spring constant *k*.

  .... **(1 mark)**

(e) Name the point on the graph that represents where the spring begins to permanently change shape.

.... **(1 mark)**

(f) Write the equation to calculate the energy stored by the spring.

.... **(1 mark)**

2 Explain the difference between the length of a spring and the extension of a spring.

....

.... **(1 mark)**

Moments

1 Identify which of the following describes a moment. Tick **one** box.

☐ A force acting in a clockwise-only direction.

☐ A perpendicular force acting at a distance from a pivot.

☐ A force acting perpendicular to a surface.

☐ A force that causes bending to occur.   **(1 mark)**

2 Describe the principle of moments for balanced objects.

When an object is balanced the clockwise moment is equal to ....

.... **(2 marks)**

3 A screwdriver is used to remove the lid from a tin of paint. Calculate the moment of the screwdriver when a force of 25 N is applied at 0.28 m from the lip of the paint lid. Select the unit from the box below.

N	N m	m

....

....

Moment = ... unit............. **(3 marks)**

4 Ben and Amberley play on a seesaw. Amberley has a mass of 25 kg and sits 1.2 m from the pivot.

Ben has a mass of 30 kg and sits 0.8 m from the pivot on the opposite side to Amberley. Take g to be 10 N/kg.

(a) Determine whether the moments that the children exert on the seesaw are in equilibrium. Write down any equation that you use.

> Remember to calculate all the quantities first so that you can use them in the correct equation. You should write down all your calculations. You need to recall the equation $M = F\,d$ because it's not on the Physics Equation Sheet.

....

....

.... **(3 marks)**

(b) Calculate the distance from the pivot Ben must sit to make the moments balance.

> The new balancing moment for Ben must equal the moment for Amberley in (a).

....

Distance = ... m **(1 mark)**

(c) Calculate how far Amberley must be from the pivot to equal the moment of Ben's original position.

....

....

Distance = ... m **(2 marks)**

Levers and gears

1 Identify which of the following statements correctly describes levers and gears. Tick **one** box.

☐ Levers and gears transmit the torsional effect of forces.

☐ Levers and gears transmit the rotational effects of forces.

☐ Levers and gears transmit the tensile effect of forces.

☐ Levers and gears transmit the compressional effect of forces.   **(1 mark)**

2 (a) Complete the diagrams of typical levers below by adding the following labels.

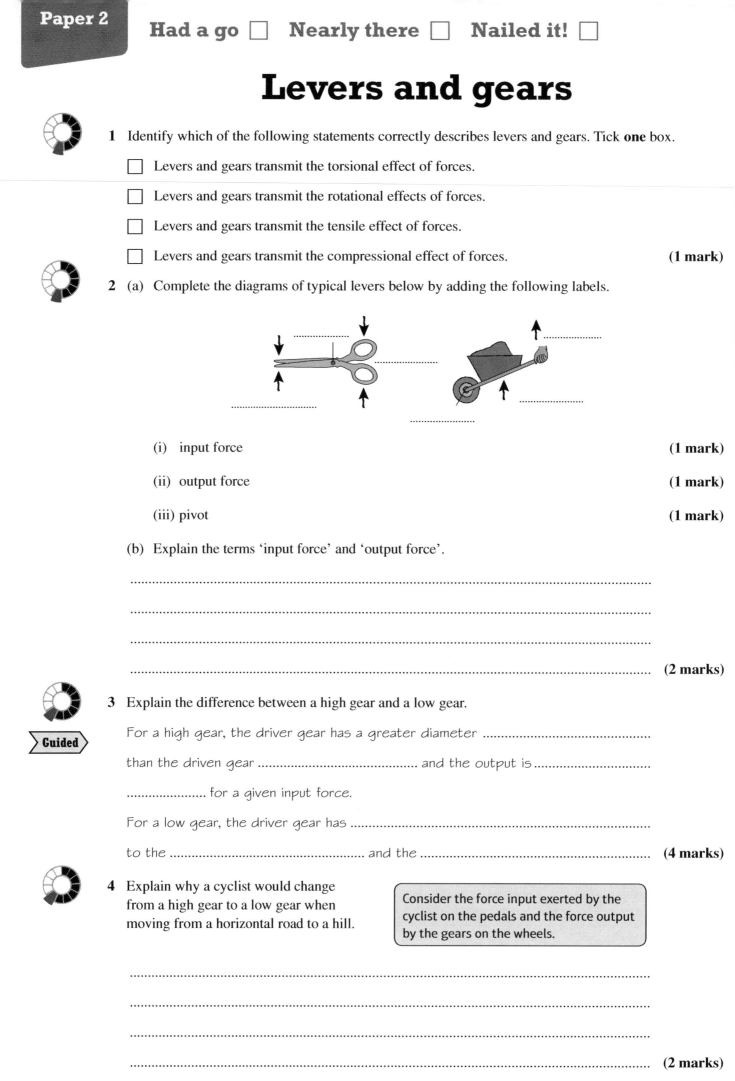

   (i)   input force   **(1 mark)**

   (ii) output force   **(1 mark)**

   (iii) pivot   **(1 mark)**

   (b) Explain the terms 'input force' and 'output force'.

....

....

....

.... **(2 marks)**

3 Explain the difference between a high gear and a low gear.

> Guided

For a high gear, the driver gear has a greater diameter ....

than the driven gear .... and the output is.............................

.... for a given input force.

For a low gear, the driver gear has ....

to the .... and the .... **(4 marks)**

4 Explain why a cyclist would change from a high gear to a low gear when moving from a horizontal road to a hill.

┌───┐
│ Consider the force input exerted by the   │
│ cyclist on the pedals and the force output │
│ by the gears on the wheels.   │
└───┘

....

....

....

.... **(2 marks)**

Pressure and upthrust

1 Choose the correct condition when an object floats in a liquid from the options below. Tick **one** box.

☐ Upthrust is more than the weight of the object.

☐ Upthrust is less than the weight of the water displaced.

☐ Upthrust is equal to the weight of the object.

☐ Upthrust is equal to the weight of the fluid displaced. **(1 mark)**

2 Calculate the pressure exerted by a rectangular block with an area of 0.0625 m² with a force of 25 N.

> You should remember $P = F / A$ because it won't be given on the Physics Equation Sheet.

....

....

Pressure = ... Pa **(2 marks)**

3 A square wooden box exerts a pressure of 2000 Pa over an area of 0.25 m². Calculate the weight of the box.

> Remember that weight is a type of force.

....

The weight of the box is found by ....

....

Weight = ... N **(2 marks)**

4 Container ships have a mark on the hull called a Plimsoll line. When this is at water level, the crew know that they must stop loading the ship. Warm seawater produces slightly less upthrust than cold seawater.

warm water   cold water

(a) Explain why the Plimsoll line on an **empty** ship will be high above the water line.

....

....

....

.... **(3 marks)**

(b) Explain why large ships that travel to warm oceans have additional Plimsoll lines on their hulls.

....

....

....

.... **(3 marks)**

Pressure in a fluid

1 Explain why a diver experiences greater pressure from the water when she swims deeper.

....

....

.... **(2 marks)**

2 (a) Calculate the water pressure on a submarine at a depth of 1500 m. Take the density of sea water to be 1025 kg/m^3 and g to be 10 N/kg.

Maths skills

> You will find this equation on the Physics Equation Sheet: $p = h\,\rho\,g$

....

....

....

Pressure = ...Pa **(3 marks)**

(b) Explain why pressure increases with depth.

....

....

.... **(2 marks)**

3 (a) Explain what is meant by atmospheric pressure.

Atmospheric pressure describes a column of air reaching from ....

Guided

....,

and covering ....

....

containing a mass of air equal to.. **(3 marks)**

(b) Calculate the mass and weight of air where an atmospheric pressure of 100 000 Pa has been recorded.

> You need to recall the equations $P = F / A$ and $F = m\,g$ as they aren't on the Physics Equation Sheet.

$P = 100\ 000$ Pa and $A = 1$ m^2 so ....

....

.... **(3 marks)**

Distance and displacement

1 The diagram shows different points on a field.

Which of these is the correct displacement from A when a footballer warms up by running from points A to B, B to D and D to C? Tick **one** box

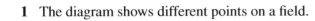

☐ A to B to D to C = 375 m

☐ AB × AC = 11 250 m

☐ A to C = 75 m

☐ AB + AC = 225 m   **(1 mark)**

2 Explain why distance is described as a scalar quantity but displacement is described as a vector quantity.

Guided

Distance does not involve .... and so is a scalar quantity.

Displacement involves both .... so it is a

vector quantity.   **(2 marks)**

3 Suzi and Sarah step into a pod on the Ferris wheel when it reaches the bottom and stops. The diameter of the Ferris wheel is 15 m. The wheel completes three cycles before the girls step off again at the bottom.

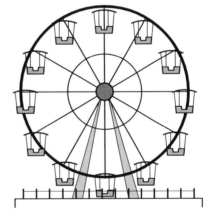

(a) Calculate the distance that the girls have travelled at the end of the ride.

Use 2 π r to calculate the circumference of a circle.

....

.... **(2 marks)**

(b) Explain the final displacement of the girls at the end of the ride.

....

.... **(2 marks)**

Speed and velocity

1 A bus travels from the bus station to the next town 10 kilometres away, passing through three villages on the way, picking up passengers at each village. The whole trip takes 40 minutes. Calculate the average speed of the bus.

....

....

> You will need to recall the equation average speed = distance / time because it's not on the Physics Equation Sheet.

Average speed = ... m/s   **(1 mark)**

2 (a) The speed of marathon runners often varies during a cross country race. Suggest **three** factors that could affect the speed of the runners.

....

.... **(3 marks)**

(b) Give typical speeds for the following:

(i) a person walking .... **(1 mark)**

(ii) a person running .... **(1 mark)**

(iii) a person cycling .... **(1 mark)**

3 A satellite orbits the Earth with a constant speed of 3100 m/s.

Explain how the speed of the satellite can be constant while the velocity is constantly changing.

Guided

The speed of the satellite is constant because the ....,

but the velocity changes constantly because the ....

.... **(2 marks)**

4 A rowing team launches the boat from the boathouse and rows up the river in a north-west direction for 15 minutes at an average velocity of 4 m/s.

> You will need to recall the equation $s = v\,t$ because it's not on the Physics Equation Sheet.

(a) Calculate the distance covered by the rowing team.

....

....

Distance = ... m   **(2 marks)**

(b) Explain why the term 'velocity' is used instead of 'speed' to describe the journey.

....

.... **(2 marks)**

(c) Describe how the displacement of the team changes as they make their way from the boathouse to finishing the journey by boat and returning to the boathouse by bus.

Guided

At the finish of the journey by boat the displacement of the team from the boathouse

is .... m.

As they return to the boathouse by bus, the displacement .... until they arrive

back at the boathouse where the final displacement is .... m.   **(3 marks)**

Distance–time graphs

1 The distance–time graph shows a runner's journey from home to the park.

   (a) Give the letter on the graph that corresponds to the part of the runner's journey where he:

       (i) stops ....

       (ii) runs fastest. ....   **(2 marks)**

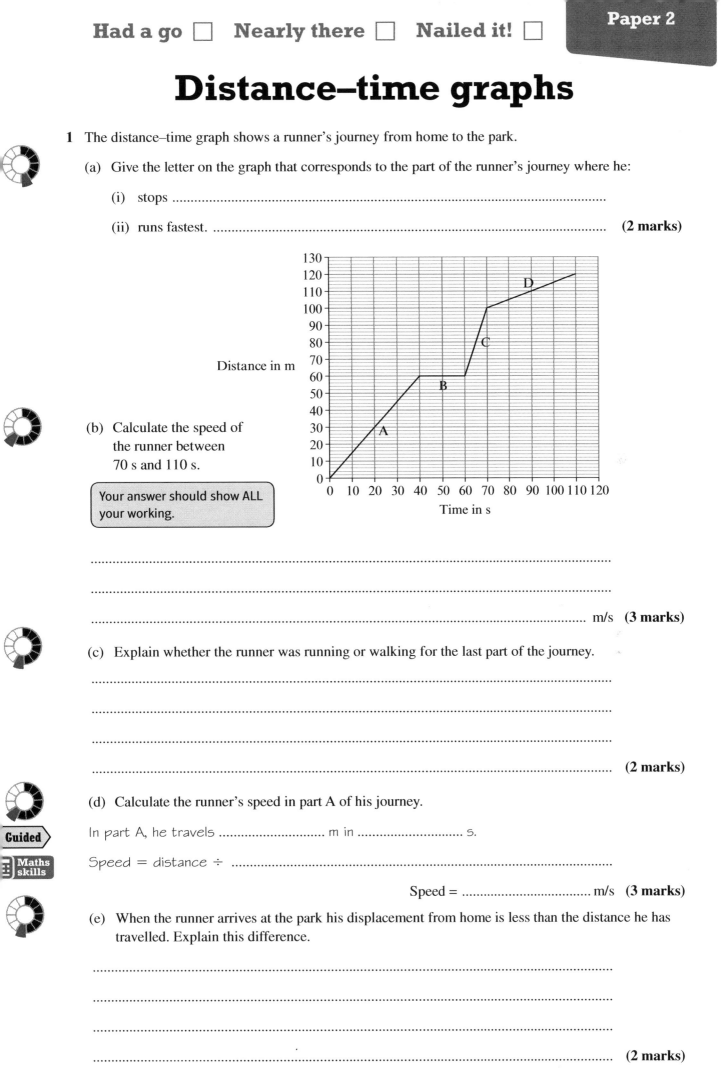

Distance in m

Time in s

   (b) Calculate the speed of the runner between 70 s and 110 s.

   > Your answer should show ALL your working.

....

....

.... m/s **(3 marks)**

   (c) Explain whether the runner was running or walking for the last part of the journey.

....

....

....

.... **(2 marks)**

   (d) Calculate the runner's speed in part A of his journey.

In part A, he travels .... m in .... s.

Speed = distance ÷ ....

                                        Speed = ... m/s **(3 marks)**

   (e) When the runner arrives at the park his displacement from home is less than the distance he has travelled. Explain this difference.

....

....

....

.... **(2 marks)**

Velocity–time graphs

1   A cyclist takes 5 seconds to reach maximum velocity of 4 m/s, from being stationary, moving in a straight line. Calculate the cyclist's acceleration. Select the unit from the box below.

Maths skills

s	m/s	m/s^2

....

....

....

Acceleration = ... unit............... **(3 marks)**

2   The velocity–time graph below shows how the velocity of a car changes with time.

Velocity in m/s

[Graph: x-axis "Time in s" from 0 to 6; y-axis "Velocity in m/s" from 0 to 35. Straight line from origin passing through the points increasing linearly.]

(a)   This graph can be used to analyse the car's journey. Identify the statement that describes the information that the graph shows. Tick **one** box.

☐ The distance the car travelled   ☐ The deceleration of the car

☐ How long the car stopped for   ☐ The constant velocity of the car

(1 mark)

(b)   Draw on the graph to show how velocity and time taken can be used to calculate acceleration.

(1 mark)

(c)   Calculate the acceleration of the car.

Guided

Change in velocity = ... m/s. Time taken for the change = ... s.

Acceleration = $\dfrac{\text{change in velocity}}{\text{time taken}}$ = ... m/s^2

Acceleration = ... m/s^2 **(2 marks)**

(d)   Use the graph to calculate the distance travelled by the car in the first 5 s.

> Calculate the area under the graph line using: ½ × base × height.

....

....

....

....

Distance = ... m   **(2 marks)**

Equations of motion

1   The velocity–time graph represents the part of a journey of a cyclist. Estimate the distance travelled by the cyclist in the first 8 seconds of the journey, using the graph.

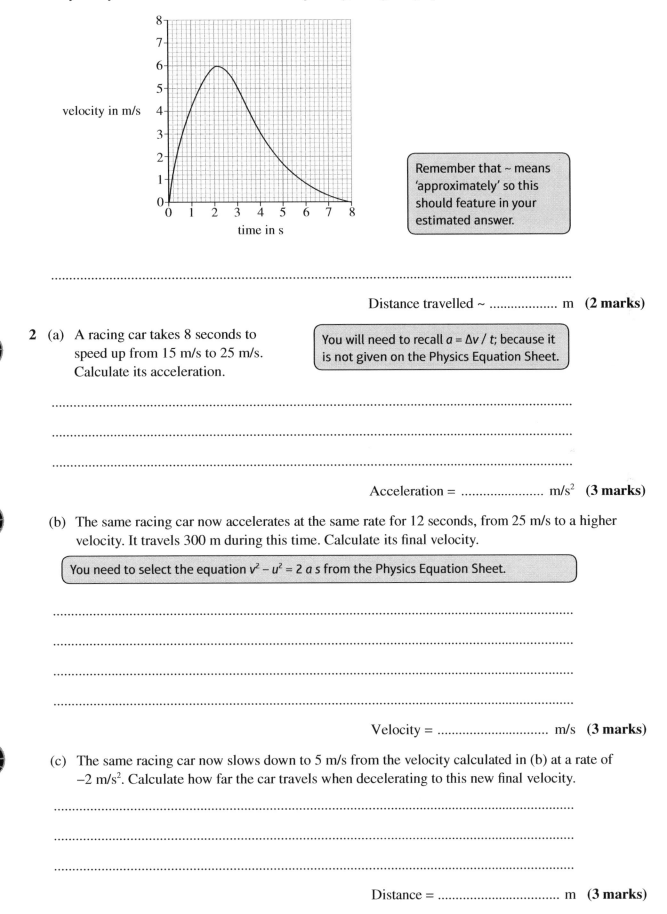

velocity in m/s

time in s

> Remember that ~ means 'approximately' so this should feature in your estimated answer.

....

Distance travelled ~ .... m   (**2 marks**)

2   (a)   A racing car takes 8 seconds to speed up from 15 m/s to 25 m/s. Calculate its acceleration.

> You will need to recall $a = \Delta v / t$; because it is not given on the Physics Equation Sheet.

....

....

....

Acceleration = ... m/s²   (**3 marks**)

(b)   The same racing car now accelerates at the same rate for 12 seconds, from 25 m/s to a higher velocity. It travels 300 m during this time. Calculate its final velocity.

> You need to select the equation $v^2 - u^2 = 2\,a\,s$ from the Physics Equation Sheet.

....

....

....

....

Velocity = ... m/s   (**3 marks**)

(c)   The same racing car now slows down to 5 m/s from the velocity calculated in (b) at a rate of −2 m/s². Calculate how far the car travels when decelerating to this new final velocity.

....

....

....

Distance = ... m   (**3 marks**)

Terminal velocity

1 An object that falls near to the Earth's surface is said to be in free fall. Identify the influence that causes the object to accelerate. Tick **one** box.

☐ Air resistance

☐ Terminal velocity

☐ Gravity

☐ Friction   **(1 mark)**

2 An experienced skydiver undergoes freefall before opening her parachute, during which she reaches a speed of around 55 m/s.

(a) Explain why the skydiver accelerates to this speed.

> **Guided**

The force of gravity pulls the skydiver downwards. ....

.... **(2 marks)**

(b) Explain why the skydiver stops accelerating at around this speed even before the parachute is opened. Refer to 'resultant force' in your answer.

> The resultant force is 0 N when two forces are in balance.

The skydiver stops accelerating because the air resistance ....

.... **(2 marks)**

(c) Give the name for the speed reached by the skydiver before she opens her parachute.

.... **(1 mark)**

(d) Explain what happens to the speed of the skydiver once the parachute is opened.

The larger surface area increases air resistance ....

....

....

....

.... **(3 marks)**

3 Ben investigates terminal velocity by dropping a ball-bearing in a very long column of viscous liquid.

Suggest how Ben could measure the terminal velocity of the ball-bearing.

> Ben will need to recall the equation $s = d / t$

....

....

....

....

....

....

.... **(4 marks)**

Newton's first law

1   A submarine is travelling at a constant depth in the sea. It starts to move forwards. Draw a free body force diagram for all the forces acting on the submarine. Label these forces.

> The length of the arrows on a free body force diagram should be proportional to the sizes of the forces.

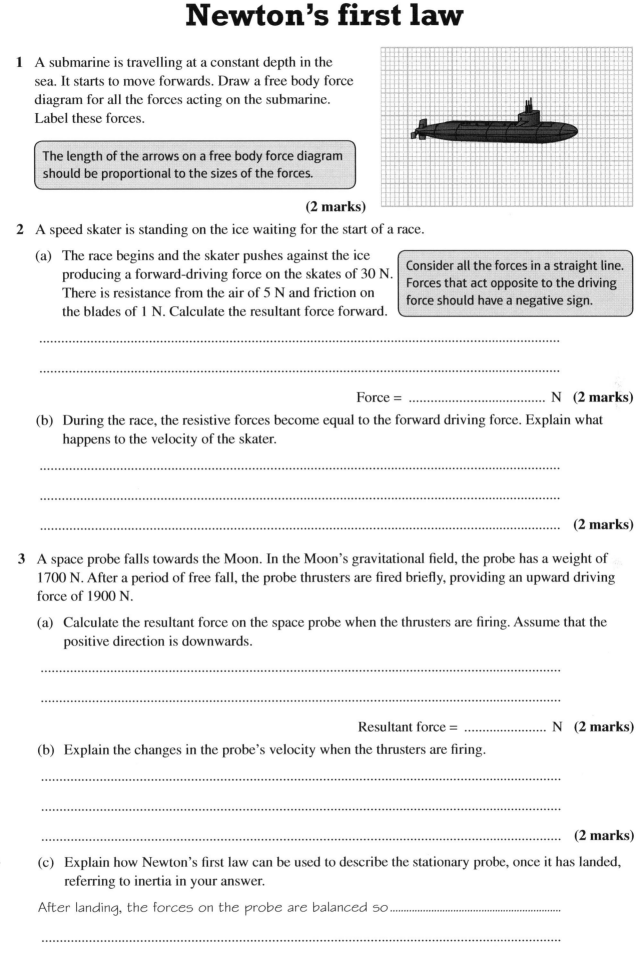

(2 marks)

2   A speed skater is standing on the ice waiting for the start of a race.

   (a)   The race begins and the skater pushes against the ice producing a forward-driving force on the skates of 30 N. There is resistance from the air of 5 N and friction on the blades of 1 N. Calculate the resultant force forward.

> Consider all the forces in a straight line. Forces that act opposite to the driving force should have a negative sign.

....

....

Force = ... N   (2 marks)

   (b)   During the race, the resistive forces become equal to the forward driving force. Explain what happens to the velocity of the skater.

....

....

.... **(2 marks)**

3   A space probe falls towards the Moon. In the Moon's gravitational field, the probe has a weight of 1700 N. After a period of free fall, the probe thrusters are fired briefly, providing an upward driving force of 1900 N.

   (a)   Calculate the resultant force on the space probe when the thrusters are firing. Assume that the positive direction is downwards.

....

....

Resultant force = ... N   (2 marks)

   (b)   Explain the changes in the probe's velocity when the thrusters are firing.

....

....

.... **(2 marks)**

Guided

   (c)   Explain how Newton's first law can be used to describe the stationary probe, once it has landed, referring to inertia in your answer.

After landing, the forces on the probe are balanced so...

....

....

.... **(3 marks)**

Newton's second law

1   In an experiment a student pulls a force meter attached to a trolley along a bench. The trolley has frictionless wheels. The force meter gives a reading of 5 N.

5N

trolley   force meter

Guided

(a) Describe what happens to the trolley.

The trolley will ....

in the direction .... **(2 marks)**

(b) The student stacks some masses on the trolley and again pulls it with a force of 5 N.

Explain why the trolley takes longer to travel the length of the bench.

The acceleration is .... because .... **(2 marks)**

Maths skills

2   A mini-bus with passengers has a mass of 3000 kg and slows down with an average acceleration of −3 m/s².

(a) Calculate the average resultant force acting on the minibus.

....

....

Force = ... N   **(2 marks)**

(b) Give the direction in which the force acts.

.... **(1 mark)**

3   A Formula One racing car has a mass of 640 kg. A resultant force of 10 500 N acts on the car.

(a) Calculate the acceleration of the racing car. Choose the unit from the box below.

m/s²	m/s	N

Remember that Newton's second law refers to $F = m\,a$

....

....

Acceleration = ... unit .... **(3 marks)**

(b) Explain what will happen to the acceleration of the car as its fuel tank empties, assuming that the resultant force remains constant and the tank never fully empties.

....

.... **(2 marks)**

Force, mass and acceleration

Practical skills

A ramp, a trolley, masses and electronic light gates can be used to investigate the relationship between force, mass and acceleration.

1 Explain **one** advantage of using electronic measuring equipment to determine acceleration compared to using a ruler and stopwatch.

....

....

.... **(2 marks)**

2 Describe the relationship between acceleration and mass.

.... **(1 mark)**

3 Explain why it is necessary to use two light gates when measuring acceleration in this experiment.

Guided

Acceleration is calculated by the change in speed / time taken, so ....

....

.... **(2 marks)**

4 (a) Describe the conclusion that can be drawn from this experiment.

Guided

For a constant slope...

....

.... **(2 marks)**

(b) Identify which of Newton's laws can be referred to in verifying the results of this experiment.

> The quantities of force, mass and acceleration are linked in this equation.

.... **(1 mark)**

5 Suggest **one** hazard associated with this experiment and **two** safety precautions that could be taken to minimise the risk of harm to the scientist.

> Consider the potential dangers of using accelerated masses or electrical equipment.

....

....

....

....

....

.... **(3 marks)**

Newton's third law

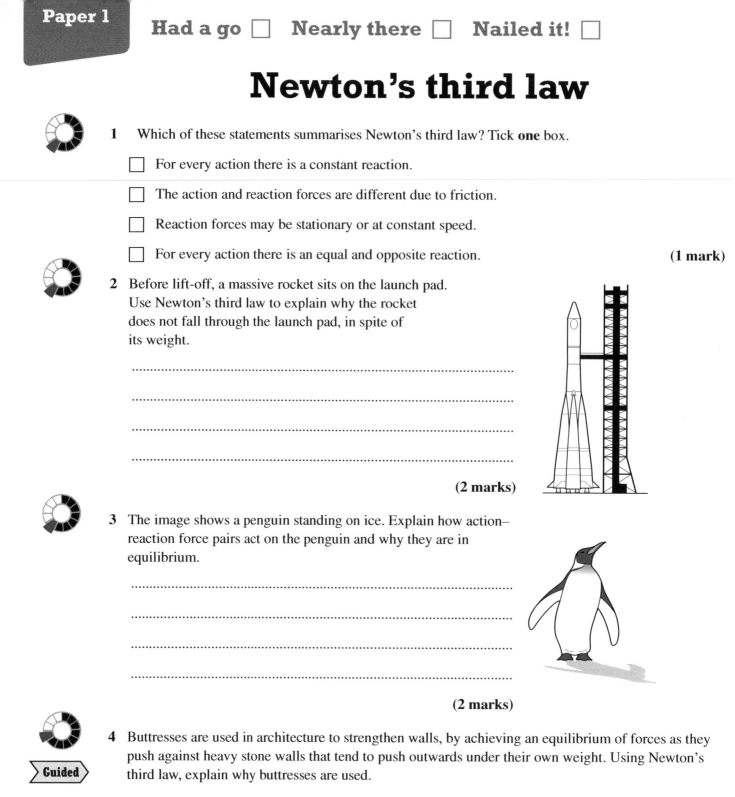

1 Which of these statements summarises Newton's third law? Tick **one** box.

☐ For every action there is a constant reaction.

☐ The action and reaction forces are different due to friction.

☐ Reaction forces may be stationary or at constant speed.

☐ For every action there is an equal and opposite reaction.

(1 mark)

2 Before lift-off, a massive rocket sits on the launch pad. Use Newton's third law to explain why the rocket does not fall through the launch pad, in spite of its weight.

....

....

....

....

(2 marks)

3 The image shows a penguin standing on ice. Explain how action–reaction force pairs act on the penguin and why they are in equilibrium.

....

....

....

....

(2 marks)

4 Buttresses are used in architecture to strengthen walls, by achieving an equilibrium of forces as they push against heavy stone walls that tend to push outwards under their own weight. Using Newton's third law, explain why buttresses are used.

Guided

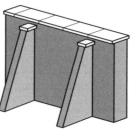

Newton's third law says that the forces must be equal in ....

and .... in direction to be in equilibrium. The force exerted by

the buttresses on the .... is equal and opposite to the

force exerted on the ....by the building, resulting in no

movement occurring.

(4 marks)

Stopping distance

1 (a) Give the word equation used to calculate overall stopping distance.

.... **(1 mark)**

(b) Calculate the overall stopping distance when a car increases its speed from 20 mph to 60 mph. Take thinking distance to be 6 m and braking distance to be 6 m when travelling at 20 mph.

....

....

....

.... **(3 marks)**

Guided

(c) Complete the table below to summarise the factors that affect overall stopping distance.

> Separate the factors that may affect the reaction time of a driver from those that affect the vehicle.

Factors increasing overall stopping distance	
Thinking distance will increase if	**Braking distance will increase if**
	the car's speed increases
the driver is distracted	

(2 marks)

Maths skills

2 Work is done on a moving car to bring it to rest. Calculate what force must be applied to the brakes of a car of mass 1500 kg travelling at 8 m/s for it to stop at the pedestrian crossing 75 m away.

....

....

....

Force = ... N **(3 marks)**

3 Recent proposals have been made to increase the national speed limit in certain cases. Suggest how these proposals might increase the risk of damage to vehicles and their passengers.

> Remember that kinetic energy is proportional to v^2.

....

....

....

....

.... **(3 marks)**

Reaction time

1 The chart below includes typical thinking distances covered by vehicles at different speeds. Identify the expected effect on overall stopping distance, if the driver was very tired. Tick **one** box.

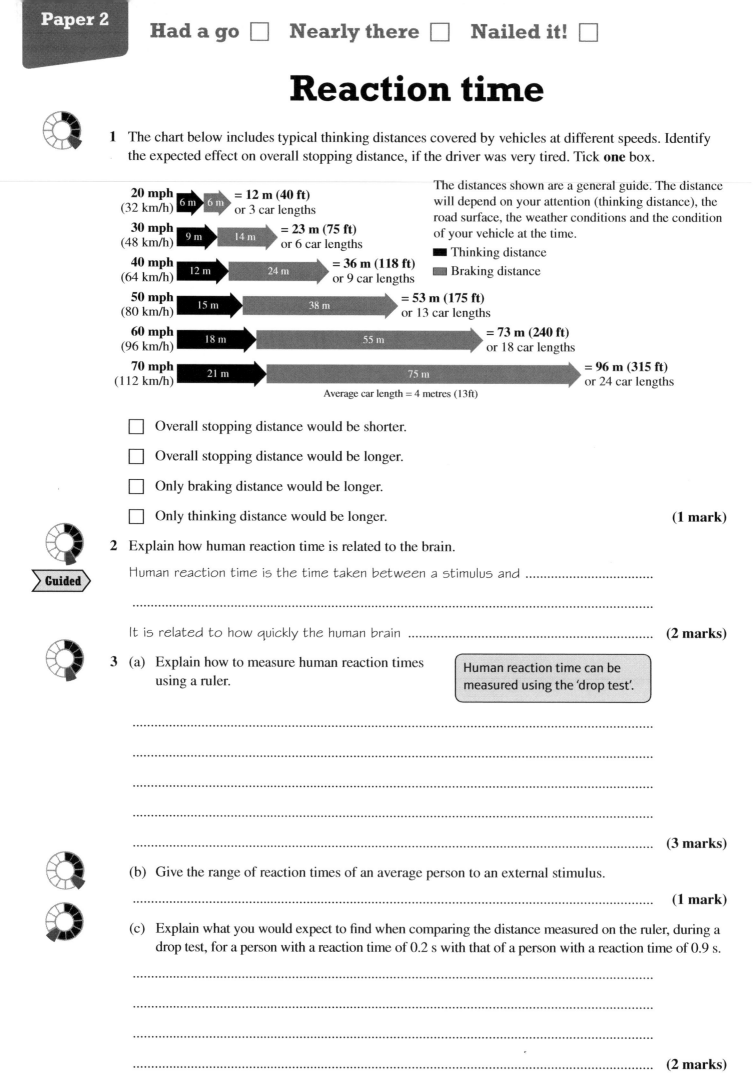

20 mph (32 km/h) 6 m 6 m = **12 m (40 ft)** or 3 car lengths

30 mph (48 km/h) 9 m 14 m = **23 m (75 ft)** or 6 car lengths

40 mph (64 km/h) 12 m 24 m = **36 m (118 ft)** or 9 car lengths

50 mph (80 km/h) 15 m 38 m = **53 m (175 ft)** or 13 car lengths

60 mph (96 km/h) 18 m 55 m = **73 m (240 ft)** or 18 car lengths

70 mph (112 km/h) 21 m 75 m = **96 m (315 ft)** or 24 car lengths

Average car length = 4 metres (13ft)

The distances shown are a general guide. The distance will depend on your attention (thinking distance), the road surface, the weather conditions and the condition of your vehicle at the time.

■ Thinking distance
■ Braking distance

☐ Overall stopping distance would be shorter.

☐ Overall stopping distance would be longer.

☐ Only braking distance would be longer.

☐ Only thinking distance would be longer.   **(1 mark)**

2 Explain how human reaction time is related to the brain.

Human reaction time is the time taken between a stimulus and ....

....

It is related to how quickly the human brain .... **(2 marks)**

3 (a) Explain how to measure human reaction times using a ruler.

> Human reaction time can be measured using the 'drop test'.

....

....

....

.... **(3 marks)**

(b) Give the range of reaction times of an average person to an external stimulus.

.... **(1 mark)**

(c) Explain what you would expect to find when comparing the distance measured on the ruler, during a drop test, for a person with a reaction time of 0.2 s with that of a person with a reaction time of 0.9 s.

....

....

....

.... **(2 marks)**

Momentum

> You will find the equation $p = m / V$ useful. You will need to learn this equation.

1 Give the **two** factors that would change the momentum of a car.

The momentum of a car would change if it .... or ....

The momentum of a car would also change if it ....

because velocity is a .... **(3 marks)**

2 Calculate the momentum of a car with a mass of 1200 kg moving at 30 m/s from north to south.

....

....

.... **(3 marks)**

3 Dima and Sam are driving dodgem cars at a funfair. The total mass of Dima and his car is 900 kg. He is moving west at 1.5 m/s.

(a) Calculate the momentum of Dima and his car in kg m/s.

....

....

Momentum = ... kg m/s **(2 marks)**

(b) Sam and his car also have a total mass of 900 kg but his car is travelling faster than Dima's car, at 3 m/s west. Sam's car collides with the back of Dima's car and both cars move forward together.

(i) Calculate the momentum of Sam and his car just **before** the collision.

....

....

Momentum = ... kg m/s **(1 mark)**

(ii) Explain what happens to the **total** of the momentum of both cars after the collision.

....

.... **(1 mark)**

(iii) Calculate the velocity of both cars as they move off together after the collision.

....

....

Velocity = ... m/s **(3 marks)**

4 A skater with a mass of 50 kg skates across the ice at 7.2 m/s in a straight line travelling north. She collides with her stationary partner who has a mass of 70 kg. They glide off together northwards. Calculate the velocity with which the pair glide across the ice.

....

....

Velocity = ... m/s **(3 marks)**

Momentum and force

1 Explain how force is related to momentum.

Guided

Force is the rate of change of.. **(1 mark)**

2 (a) A car with a mass of 1500 kg travels at 25 m/s along a motorway. It crashes into a central barrier and stops in 1.8 seconds, resulting in a momentum of zero. Calculate the initial momentum of the car.

> You will find the equation $p = m\,v$ useful.
> You will need to learn this equation.

....

....

....

....

Momentum = ... kg m/s   **(3 marks)**

(b) Calculate the force acting on the car at the time of impact.

> You will find the $F = m\,\Delta v / \Delta t$ equation useful.
> You will need to select this equation from the Physics Equation Sheet.

....

....

....

....

Force = ... N   **(2 marks)**

3 (a) Explain how a large force is exerted on a passenger in a vehicle in the event of a crash.

Guided

The forces exerted on the passenger are large when the .... or the

.... of the vehicle are large.   **(2 marks)**

(b) Suggest how the force on the passengers in the event of a crash can be reduced.

An .... increases the time over which a passenger

comes to rest so ....   **(2 marks)**

4 Calculate the force on a motorbike of mass 500 kg as it speeds up from 10 m/s to 15 m/s in 20 s.

Maths skills

....

....

....

Force = ... N   **(2 marks)**

Extended response – Forces

A student investigates circular motion by tying a 57 g tennis ball to a string which is then rotated in a horizontal plane at constant speed. The student counts the number of rotations.

Explain how acceleration and centripetal force are considered in this experiment.

Your answer should include an example of how the experiment may be extended to improve data collection.

> You will be more successful in extended response questions if you plan your answer before you start writing.
>
> The question asks you to give a detailed explanation of acceleration and centripetal force. Think about:
>
> - Why the tennis ball is described as accelerating.
>
> - How centripetal force is described.
>
> - How the student can improve his investigation by changing variables.
>
> - How the student can improve data collection.
>
> - Identify the importance of control variables.
>
> You should try to use the information given in the question.

....

....

....

....

....

....

....

....

....

....

....

....

....

....

....

.... **(6 marks)**

Waves

1 Describe evidence that it is energy in a wave that travels, not the particles themselves, for ripples on the surface of water.

....

....

.... **(2 marks)**

2 Explain how a loudspeaker can generate a sound wave in air.

> Sound is generated due to the vibrations of the moving cone in the loudspeaker.

....

....

.... **(2 marks)**

3 The diagram below shows a wave travelling through a medium.

Height of wave in cm

> Note that both axes have units of cm.

Distance from source in cm

(a) Identify the correct amplitude of the wave in the diagram. Tick **one** box.

☐ 0.05 m   ☐ 0.025 m   ☐ 0.12 m   ☐ 0.10 m   **(1 mark)**

(b) Determine the wavelength of the wave in the diagram.

....

Wavelength = ... m   **(1 mark)**

(c) Sketch a second wave on the diagram to show a higher amplitude and shorter wavelength than that of the wave shown.   **(2 marks)**

(d) Determine the time period T of the wave if the speed of the wave is 3×10^8 m/s.

> You will need to recall the equation $v = f\lambda$ as it is not on the Physics Equation Sheet.
> You will need to be able to select the equation $period = 1/frequency$ from the Physics Equation Sheet.

> **Guided**

First find the frequency of the wave ....

.... **(2 marks)**

Then find the time period of the wave ....

.... **(2 marks)**

Wave equation

1   Blue whales communicate over long distances by sending sound waves through the oceans.

(a) If the speed of the sound through water is 1500 m/s, calculate the frequency of one note with a wavelength of 88 m produced by a blue whale.

Using $v = f\lambda$ rearrange the equation to give f = ...

....

Frequency = ... Hz **(2 marks)**

(b) If the note sent from another whale had a different frequency of 22 Hz, show that the wavelength of this note would also change from the wavelength in (a).

....

Wavelength = ... m **(2 marks)**

2   A sound wave has a wavelength of 0.017 m and a frequency of 20 000 Hz.

Calculate the speed of the wave.

....

Wave speed = ... m/s **(2 marks)**

3   An icicle is melting into a pool of water. Drops fall every half a second, producing small waves that travel across the water at 0.05 m/s. Calculate the wavelength of the small waves. Select the correct unit from the box below.

seconds	hertz	metres

Remember to write down the equation that you are using before you substitute values.

....

....

Wavelength = ... unit .... **(3 marks)**

4   A satellite orbiting at 36 000 km above the Earth sends signals to a satellite dish on the ground using radio waves. It takes 0.12 s for the radio waves to travel from the satellite to the dish. The radio waves have a wavelength of 5 m.

Calculate the frequency of the radio waves.

You will need to do this calculation in two steps, first using the equation $s = d / t$

....

....

....

....

Frequency = ... Hz **(4 marks)**

Measuring wave velocity

> **Guided**

Maths skills

1   A tap is dripping into a bath. Three drops fall each second, producing small waves that are 5 cm apart.

Calculate the speed of the small waves across the water. Select the correct unit from the box below.

metres per second² (m/s²)	metres per second (m/s)	metres (m)

The frequency of the waves (*f*) = ...

The wavelength of the waves (λ) = ...

....

Speed of waves = ... unit .... **(3 marks)**

Maths skills

2   Identify the distance between the crests of water waves with a frequency of 0.25 Hz travelling at a speed of 2 m/s. Tick **one** box.

> You will find the equation *v = f* λ useful.
> You will need to select this equation from the Physics Equation Sheet.

☐ 0.5 m

☐ 0.125 m

☐ 4 m

☐ 8 m   **(1 mark)**

3   An oscilloscope screen shows a waveform.

> You will find the equation
> *period = 1 / frequency* useful.
> You will need to select this equation from the Physics Equation Sheet.

Maths skills

Each division in the horizontal direction is 5 milliseconds. Calculate the frequency of the wave.

....

....

....

Frequency = ... Hz **(3 marks)**

Practical skills

Waves in fluids

1 A ripple tank is used to investigate waves.

(a) Describe how a ripple tank may be used to measure the frequency of water waves.

....

....

.... **(2 marks)**

(b) Describe how to find the wavelength of the waves in the ripple tank.

....

....

.... **(2 marks)**

(c) Give the equation you can use with the data collected in (a) and (b) to determine wave speed.

.... **(1 mark)**

(d) Identify the control variable when using a ripple tank to investigate wave speed.

.... **(1 mark)**

2 Describe a suitable conclusion to the method of using the ripple tank in Q1. Your conclusion should include two factors that should be moderated in this experiment.

Guided

A ripple tank can be used to determine a value for ....

....

.... **(3 marks)**

3 The ripple tank experiment uses several pieces of equipment. Complete the table below to describe the hazard associated with each component and suggest a measure to minimise the risk of harm.

> Identify the hazard and describe the safety measure for each mark.

Component	Hazard	Safety measure
water		
electricity		
strobe lamp		

(3 marks)

Waves and boundaries

1 (a) Identify which of the following will reflect the greatest amount of sound energy in air. Tick **one** box.

☐ Hedge ☐ Stone wall

☐ Wooden fence ☐ Curtain **(1 mark)**

(b) Explain your answer to (a).

....

.... **(2 marks)**

Guided

2 Describe how a periscope works by referring to the simplified diagram. Include the following terms in your answer: angle of incidence, plane, angle of reflection, perpendicular, normal.

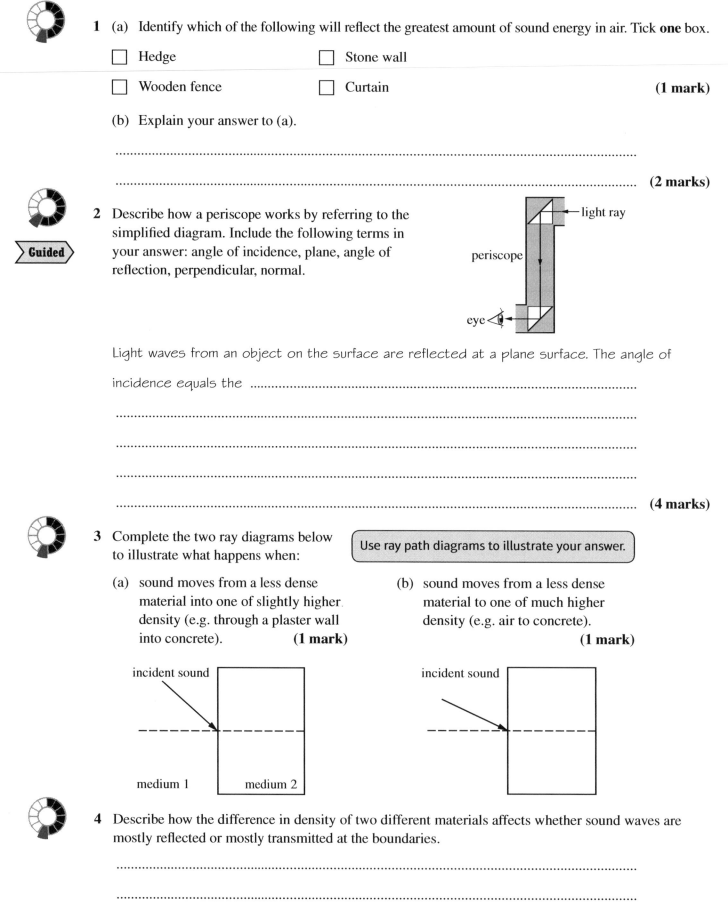

Light waves from an object on the surface are reflected at a plane surface. The angle of

incidence equals the ....

....

....

....

.... **(4 marks)**

3 Complete the two ray diagrams below to illustrate what happens when:

> Use ray path diagrams to illustrate your answer.

(a) sound moves from a less dense material into one of slightly higher density (e.g. through a plaster wall into concrete). **(1 mark)**

(b) sound moves from a less dense material to one of much higher density (e.g. air to concrete). **(1 mark)**

incident sound

medium 1   medium 2

incident sound

4 Describe how the difference in density of two different materials affects whether sound waves are mostly reflected or mostly transmitted at the boundaries.

....

....

....

.... **(3 marks)**

Investigating refraction

1 (a) Suggest a method that could be used to investigate the refraction of light using a glass block and a ray box.

....

....

....

....

....

....

.... **(4 marks)**

(b) Explain what conclusion you would expect to find using the method you have outlined in (a). Your answer should refer to the angle of incidence and the angle of refraction.

⟩**Guided**⟩ When a light ray travels from air into a glass block ....

....

.... **(2 marks)**

(c) (i) Explain what would be observed if the light ray, travelling through the air, entered the glass at an angle of 90° to the surface of the glass.

.... **(1 mark)**

(ii) Explain what would be happening that could not be observed in this experiment.

....

.... **(2 marks)**

2 Give **three** hazards and appropriate safety measures that should be considered when investigating reflection using a mains powered ray box.

....

....

....

....

.... **(3 marks)**

3 The refraction of light waves through transparent materials can be modelled using a ripple tank.

Describe what changes you would expect to observe in the waves when the depth of the water is made shallower by placing a glass sheet at an angle to the waves.

> Consider the waves being generated in deeper water and then moving into shallower water.

....

....

....

.... **(2 marks)**

Sound waves and the ear

1   Identify the correct sequence of sound waves in air when a guitar is played. Tick **one** box.

(a)   Energy travels via longitudinal waves through the air.

(b)   The air molecules near the guitar string vibrate in response to movement of the string.

(c)   The eardrum vibrates.

(d)   Longitudinal waves are channelled into the ear canal.

☐ a, c, d, b   ☐ c, d, b, a

☐ b, a, d, c   ☐ a, b, c,   **(1 mark)**

2   Explain why sound travels more slowly in air than in water.

....

....

....   **(2 marks)**

3   (a)   Explain how sound from a tuning fork is
transmitted to the fluid in the inner ear.

> Sound is transmitted by pressure
> waves travelling through a range of
> materials to reach the inner ear.

external ear (pinna)

eardrum

sound waves

ear canal

....

....

....

....

....   **(3 marks)**

(b)   Explain how vibration in the inner ear is sent to the brain.

....

....

....

....   **(2 marks)**

4   Some animals communicate using frequencies below 20 Hz. Others use frequencies over 20 kHz.
Explain why humans cannot hear these sounds.

Guided

In the human ear, the eardrum will not vibrate if the wave frequency is less than ....

....

....

....

....

....   **(3 marks)**

Uses of waves

1 Ultrasound can be used by oceanographers to explore the seabed. An ultrasound pulse travels through seawater at 1500 m/s and an echo is heard 4 seconds after transmission. Choose which calculation can be used to determine the correct depth of the ocean at that point. Tick **one** box.

> Remember that the sound wave needs to travel to and from the seabed.

☐ (1500 / 4) = 375 m   ☐ (1500 × 4) / 2 = 3000 m

☐ (1500 × 4) / 4 = 1500 m   ☐ (1500 × 4) = 6000 m   **(1 mark)**

2 Describe how the reflection of ultrasound is used to make a picture of a fetus in the womb.

As ultrasound waves pass into the body, some waves are reflected ....

....

....

.... **(4 marks)**

3 The structure of the Earth can be analysed using information from seismic waves. It has been found that the speed of seismic waves from the solid crust increases as they go deeper into the Earth's semi-solid mantle.

(a) Suggest an explanation for this increase in speed.

....

.... **(3 marks)**

(b) Calculate the wavelength of seismic waves generated by an earthquake. The waves travel at 7 m/s with a frequency of 0.05 Hz.

....

....

Wavelength = ... m **(2 marks)**

4 Ultrasound is used to produce images of the structure of computer chips because the waves are reflected by the different layers of materials. An ultrasound signal is sent into the top surface of a chip and an echo is detected from a layer deeper in the chip after 0.5 nanoseconds. The speed of sound in the computer chip is 8400 m/s. Calculate the distance of the layer from the surface of the chip.

> You will need to answer this question in two steps. First calculate total distance travelled by the ultrasound, then calculate the distance to the layer that reflects the ultrasound.

....

....

....

....

....

Distance = ... m **(4 marks)**

Electromagnetic spectrum

1 Visible and infrared radiation are given out by a candle. Gamma-rays are emitted by radioactive elements such as radium.

(a) Give **two** similarities between all waves in the electromagnetic spectrum.

All waves of the electromagnetic spectrum are .... waves

and they all travel .... **(2 marks)**

(b) Explain how energy is related to electromagnetic waves.

Electromagnetic waves all transfer energy from ....

.... **(1 mark)**

2 The chart below represents the electromagnetic spectrum. Some types of electromagnetic radiation have been labelled.

(a) Name the three parts of the spectrum that have been replaced by letters in the diagram.

longest wavelength/ lowest frequency   shortest wavelength/ highest frequency

←——— radio waves ———→◄—C—►◄infrared►B   ◄—►◄——A——→
                                     ultra-
                                     violet   ◄gamma►
                                     rays (UV)   rays

A: ....

B: ....

C: .... **(3 marks)**

(b) Describe how frequency changes from radio to gamma waves and how this is related to energy. Refer to these waves in your answer.

....

....

....

....

.... **(3 marks)**

3 The speed of electromagnetic waves in a vacuum is 300 000 km/s. A radio wave has a wavelength of 240 m. Calculate the frequency of the radio wave.

> You will need to remember the equation $v = f\lambda$ because it's not on the Physics Equation Sheet.

....

....

....

....

....

Frequency = ... Hz   **(3 marks)**

Properties of electromagnetic waves

1 Which of the following is **not** true about the behaviour of electromagnetic waves? Tick **one** box.

☐ Electromagnetic waves travel at 300 000 000 m/s in a vacuum.

☐ Electromagnetic waves are transverse waves.

☐ Electromagnetic waves are all transmitted from space through the atmosphere.

☐ Electromagnetic waves change speed in different materials.   **(1 mark)**

2 (a) Describe **four** properties of electromagnetic waves that explain their behaviour.

> Important properties of waves are reflection, refraction, transmission and absorption.

....

....

....

....

....

.... **(4 marks)**

(b) Give examples of **two** of the properties described in (a).

....

.... **(2 marks)**

3 (a) Explain the differences between microwaves and radio waves in terms of frequency and wavelength.

....

.... **(2 marks)**

(b) Describe how the properties of microwaves and radio waves dictate the way they are transmitted in communications systems.

> Consider how the ionosphere affects the route of the waves from transmitter to receiver.

Guided

Microwaves sent from the transmitter ....

and are received and re-emitted ....

Radio waves are sent from the transmitter but are ....

and then .... **(4 marks)**

4 Explain why space-based telescopes have been able to collect extra information about the Universe, beyond that collected by terrestrial telescopes based on Earth.

....

....

.... **(3 marks)**

Had a go ☐ Nearly there ☐ Nailed it! ☐

Infrared radiation

1 (a) Describe an experimental method, using the apparatus in the diagram below, to investigate the radiation of thermal energy.

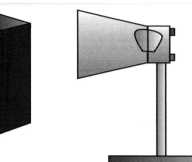

> The diagram shows a Leslie's Cube which has four different sides – black, white, shiny and dull, and can be filled with hot liquid.

....

....

....

.... **(4 marks)**

Guided

(b) Identify the independent and dependent variables in this experiment.

The dependent variable is temperature.

The independent variable is...

.... **(2 marks)**

(c) Name **four** control variables in this experiment.

....

.... **(2 marks)**

2 (a) Identify a hazard with using Leslie's cube.

.... **(1 mark)**

(b) Explain how the hazard can be minimised.

.... **(1 mark)**

3 Conical flasks covered with different coloured foils and containing cold water may be used to investigate the absorption of thermal energy, by heating them with a radiant heater placed at a measured distance from the flasks.

(a) Suggest a reason why the conical flasks should be fitted with bungs before starting the experiment.

.... **(1 mark)**

(b) Describe what you would expect to find at the end of the experiment for:

(i) dull and black surfaces

.... **(1 mark)**

(ii) shiny and light surfaces.

.... **(1 mark)**

Dangers and uses

1 Give **two** uses for each of the following EM waves.

(a) Infrared waves:

1 .... 2 .... **(1 mark)**

(b) Ultraviolet waves:

1 .... 2 .... **(1 mark)**

(c) Gamma waves:

1 .... 2 .... **(1 mark)**

(d) Give a use for microwaves other than for cooking using a microwave oven.

.... **(1 mark)**

2 Medical X-rays are used to diagnose certain types of medical problems. The numbers of X-rays are carefully recorded.

Medical X-rays	Adult approximate effective radiation dose	Equivalent time of exposure to background radiation
radiography – chest	0.1 mSv	10 days
oral	0.005 mSv	1 day
computed tomography (CT) – lung cancer screening	1.5 mSv	6 months

(a) Using data from the table, compare the medical X-ray doses with background radiation.

....

....

....

....

....

.... **(3 marks)**

(b) Suggest why it is important to record and control the number of X-rays that a person has.

X-rays carry high amounts of energy.

....

....

....

.... **(2 marks)**

3 Explain how radio waves may be produced by, and induced in, electrical circuits.

Guided

Radio waves can be produced by ....

Radio waves can also .... in electrical circuits by

....

creating an .... with the same frequency

as the radio waves, when they are ....

.... **(4 marks)**

Lenses

1   Describe the main difference between the way converging and diverging lenses bend light.

A converging lens bends the rays of light towards each other ....

A diverging lens bends the .... **(2 marks)**

2   (a)   Explain how the focal point of a lens is related to its thickness.

....

.... **(2 marks)**

     (b)   Describe how the focal length of a lens is identified.

.... **(1 mark)**

3   The diagram below shows two converging lenses.

     (a)   Complete the diagrams for both lenses by drawing the rays to converge at a focal point.

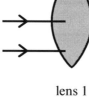

lens 1

lens 2   **(4 marks)**

     (b)   Name the type of lens shown in the diagram.

.... **(1 mark)**

4   The symbol below represents a concave lens.

> Remember that concave and convex lenses are represented by symbols.

     (a)   Draw three rays to show how light passes through the lens, label the focal point and the focal length.   **(4 marks)**

     (b)   (i)   Name the type of image you would expect to be formed by the concave lens.

.... **(1 mark)**

          (ii)   Explain the reasons for your answer to (b)(i).

....

.... **(2 marks)**

Real and virtual images

1 Explain the difference between a real image and a virtual image.

A real image is an image that can be projected ....

but a virtual image ....

A virtual image is produced when .... **(3 marks)**

2 Add the following points to the diagram:

You will need to use your ruler.

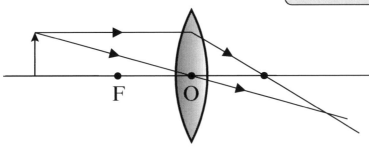

F   O

(a) 2F on each side of the lens   **(1 mark)**

(b) The second focal point   **(1 mark)**

(c) Arrow showing the position and type of image   **(1 mark)**

3 A fossil measuring 3.3 mm is examined under a magnifying lens with a magnification of 4.5. Calculate the height of the resultant image of the fossil. Use the correct equation from the Physics Equation Sheet. Show your calculations.

You will find this equation useful:
$$\text{magnification} = \frac{\text{image height}}{\text{object height}}$$

....

....

.... **(2 marks)**

4 The table below shows conditions for a converging lens.

Position of object	Position of image	Type of image	Magnification	Upright/inverted
closer than F	in front of lens	virtual	magnified	upright
between F and 2F	beyond 2F on the opposite side of the lens	real	magnified	inverted
at 2F	at 2F on the opposite side of the lens	real	same size	inverted
beyond 2F	F to 2F on the opposite side of the lens	real	smaller	inverted

(a) Describe what happens to the magnification of the image as the object gets closer to the focal point from 2F.

.... **(1 mark)**

(b) Give the point at which the image changes to a virtual image.

.... **(1 mark)**

(c) Give the point at which the image is real and the same size as the object.

.... **(1 mark)**

Visible light

1 Identify the correct categories of materials that transmit light. Tick **one** box.

☐ Opaque and transparent

☐ Opaque and translucent

☐ Translucent and transparent

☐ Transparent only   **(1 mark)**

2 Give an example of where specular reflection occurs.

.... **(1 mark)**

3 Explain why an opaque green object appears green in white light.

> **Guided**

An opaque green object appears green because ....

....

and all other colours .... **(2 marks)**

4 Suggest a scientific principle to explain the following statements about coloured light.

(a) Violet light carries more energy than red light.

....

.... **(1 mark)**

(b) A green book cover with a blue title under white light, is seen as a green book cover with a black title when placed under yellow light.

....

....

.... **(3 marks)**

5 Explain, with the aid of a diagram, how diffuse reflection still obeys the law of reflection even though light is scattered.

> Remember that reflection from a rough surface causes scattering.

....

....

....

....

.... **(5 marks)**

Black body radiation

1 A hot cup of coffee is left on the table and cools down. Which statement explains why this happens? Tick **one** box.

☐ The cup absorbs more radiation than it emits.
☐ The cup emits more radiation than it absorbs.
☐ The cup emits the same amount of radiation as it absorbs.
☐ Radiation is transferred from the surroundings.

(1 mark)

2 Two cups of coffee are left at the same time to cool down. The starting temperature for Cup A is 80 °C and the starting temperature for Cup B is 55 °C. Explain which cup will radiate the most heat in 5 minutes.

....

....

.... **(2 marks)**

3 Carbon dioxide absorbs thermal energy and is described as a greenhouse gas. Describe the main factors affecting the temperature of the Earth and how an increase in atmospheric carbon dioxide may lead to an increase in global warming.

> Thermal equilibrium relies on a balance between absorption and emission of energy.

Guided

Maintenance of the average temperature of the Earth relies on a balance between

absorption of energy from...

....

....

....

.... **(4 marks)**

4 Astronomers can analyse the temperature of stars by examining the wavelengths and intensity of light that the stars emit. Describe how astronomers can use these data to determine the temperature of a star.

....

....

....

....

....

.... **(2 marks)**

93

Extended response – Waves

X-rays and gamma-rays are widely used in a number of applications. Compare these waves and give examples of how they can be used safely in industry.

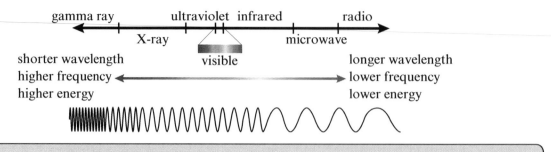

You will be more successful in extended response questions if you plan your answer before you start writing.

The question asks you to give a detailed explanation of the properties, uses and dangers of X-rays and gamma waves. Think about:

- The types of waves and how you would describe them.
- The dangers of both types of waves and the reasons why they can be dangerous.
- Examples of how the waves are used in medicine.
- Examples of how the waves are used in industry.

You should try to use the information given in the question and in the diagram.

....

....

....

....

....

....

....

....

....

....

....

....

....

....

....

.... **(6 marks)**

Magnets and magnetic fields

1 Complete the diagram below to show the magnetic field lines for a bar magnet.

> Remember to consider the relative density of the field lines as well as their polarity.

(4 marks)

2 Write down **three** similarities between the magnetic field of a bar magnet and that produced by the Earth.

> **Guided**

A bar magnet and the Earth both have ....

A bar magnet and the Earth have similar ....

The direction of both fields can be found using .... **(3 marks)**

3 An electric doorbell uses an induced magnet to move the hammer, which then rings the bell when the switch is pressed.

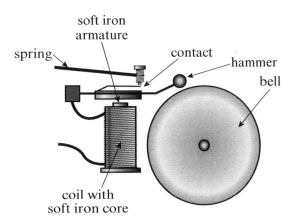

Explain why an induced magnet rather than a permanent magnet is used for this application.

....

....

....

....

....

.... **(5 marks)**

Had a go ☐ Nearly there ☐ Nailed it! ☐

Current and magnetism

1 The diagrams below show a wire passing through a circular card. The **cross** represents the conventional current moving into the page and the **dot** represents the conventional current moving out of the page.

⊙

⊗

(a)

(b)

(a) Draw magnetic field lines on each diagram to show the **pattern** of the magnetic field. **(2 marks)**

(b) Draw arrows on each diagram to show the **direction** of the magnetic field. **(2 marks)**

2 Identify which of the following has a magnetic field similar in shape to that of a solenoid. Tick **one** box.

☐ Ball magnet

☐ Bar magnet

☐ Circular magnet

☐ Horseshoe magnet **(1 mark)**

> **Guided**

3 Give **three** factors that affect the force acting on a current-carrying wire in a magnetic field.

The force acting on a current-carrying wire in a magnetic field depends on the ......

.... of the wire, the .... in the wire and the

.... **(3 marks)**

Maths skills

4 In an experiment, Alaric takes measurements to calculate the force acting on a long straight conductor of length 56 cm when it is placed in a magnetic field of flux density 0.005 T. The current flowing in the wire is 1.4 A.

> You will need to select this equation from the Physics Equation Sheet: $F = B\,I\,l$

(a) Calculate the force acting on the wire.

....

$F =$ .... N **(2 marks)**

(b) Calculate the force acting when the current is increased to 2.8 A.

....

$F =$ .... N **(2 marks)**

(c) Calculate the force acting when the current is 1.4 A but length is reduced to 23 cm.

....

$F =$ .... N **(2 marks)**

Current, magnetism and force

1 (a) Explain the effect that you would expect to observe when a current-carrying wire is placed in a magnetic field. Use the diagrams to help you.

wire carrying current *I*

N   S

When a current-carrying wire is placed in a magnetic field it would ....

This is because the current-carrying wire has ....

which interacts with the ....   **(3 marks)**

(b) Identify the correct term for the effect produced when the magnetic fields of a magnet and a current-carrying conductor exert a force on each other. Tick **one** box.

☐ Alternating effect   ☐ Induction effect

☐ Generator effect   ☐ Motor effect   **(1 mark)**

2 Label the diagram illustrating Fleming's left hand rule to show how the **direction** of the force acting on a current-carrying wire in a magnetic field may be determined.

....

....

....

(3 marks)

3 Explain how the size of the force acting on a current-carrying wire in a magnetic field may be increased.

The size of the force can be increased by increasing the strength of the ....

....

or by ....   **(2 marks)**

4 A wire of length 30 cm is placed in a magnetic field of flux density 0.0005 T and is acted on by a force of 0.21×10^{-3} N. Calculate the current carried in the wire. Use the correct equation from the Physics Equation Sheet. Choose the correct unit from the box.

| newtons / N | tesla / T | amps / A |

You need to use the equation $F = B I l$

....

....

....

Current = ... unit .... **(3 marks)**

The motor effect

1 The magnetic field produced in a current-carrying wire is weak. Suggest how the field produced from the wire may be made stronger.

.... **(1 mark)**

2 (a) The diagram shows a generator. Describe **two** methods of changing the direction of rotation of the coil when generating an a.c. current.

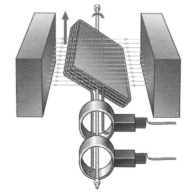

> Your answer should mention current and magnetic flux.

....

....

....

.... **(2 marks)**

(b) Suggest an application where it may be useful to reverse the direction of a motor.

.... **(1 mark)**

3 Suggest **three** ways that the speed of rotation of a motor may be increased.

1. ....

2. ....

3. .... **(3 marks)**

4 Explain the importance of the split-ring commutator in the operation of a d.c. motor.

> Guided

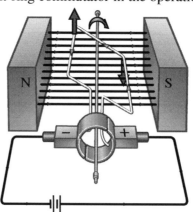

The commutator is in contact with .... but it is split into two parts

creating ....

As the motor spins, the contacts touch ....

causing the current to .... **(4 marks)**

Induced potential

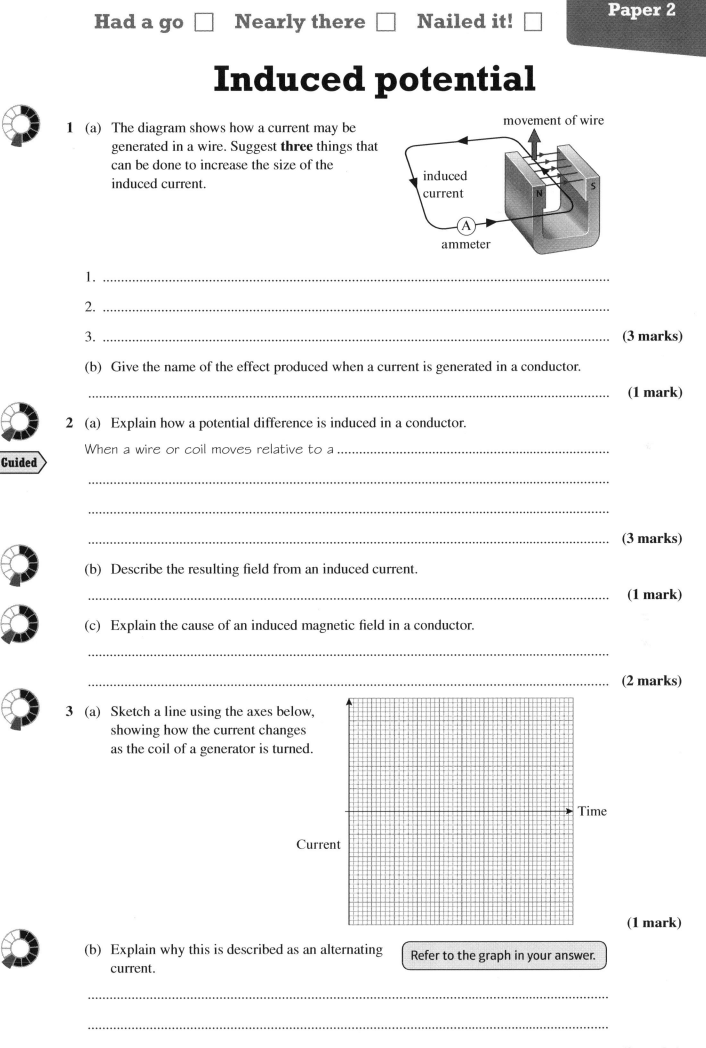

1 (a) The diagram shows how a current may be generated in a wire. Suggest **three** things that can be done to increase the size of the induced current.

movement of wire

induced current

A

ammeter

N   S

1. ....

2. ....

3. .... **(3 marks)**

(b) Give the name of the effect produced when a current is generated in a conductor.

.... **(1 mark)**

2 (a) Explain how a potential difference is induced in a conductor.

When a wire or coil moves relative to a ....

....

....

.... **(3 marks)**

(b) Describe the resulting field from an induced current.

.... **(1 mark)**

(c) Explain the cause of an induced magnetic field in a conductor.

....

.... **(2 marks)**

3 (a) Sketch a line using the axes below, showing how the current changes as the coil of a generator is turned.

Time

Current

(1 mark)

(b) Explain why this is described as an alternating current.

Refer to the graph in your answer.

....

....

.... **(2 marks)**

Alternators and dynamos

1 Explain the terms alternating current and direct current.

....

....

....

.... **(2 marks)**

2 Identify the correct pair of statements about alternators and dynamos. Tick **one** box.

	Alternator	Dynamo
☐	produces a d.c. output	produces an a.c. output.
☐	slip rings produce a d.c. current	split-ring commutator produces an a.c. current
☐	slip rings produce an a.c. current	slip rings produce a d.c. current
☐	produces an a.c. output	produces a d.c. output.

(1 mark)

3 The graphs below show how current varies with time for two types of power generator: an alternator and a dynamo.

(a) Name the type of generator that was used to produce each graph.

Current (*I*)   A   Time in s   Current (*I*)   B   Time in s

A .... **(1 mark)**

B .... **(1 mark)**

(b) Explain your answers to (a) by referring to the graphs.

    (i) Graph A shows that at every half ....

.... **(1 mark)**

    (ii) Graph B shows that at every half...

.... **(1 mark)**

4 Discuss the similarities and differences between an alternator and a dynamo, in terms of how they produce current.

> Consider the influence of the magnetic fields in the process.

An alternator and a dynamo both use the interaction of ....

to produce .... As the coil in the alternator rotates,

the way it faces is continually changing creating an ....

....

but, as the coil in a dynamo rotates, the way it faces also changes but, as the contacts

.... a .... is produced. **(5 marks)**

Loudspeakers

1 Identify the conversion involved in a working loudspeaker. Tick **one** box.

☐ Electrical current variations into pressure sound waves

☐ Magnetic force into electrical current variations

☐ Pressure sound waves into electrical current variations

☐ Pressure sound waves into magnetic force **(1 mark)**

2 Explain how loudspeakers use the motor effect.

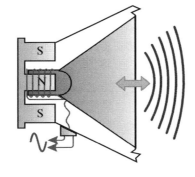

> **Guided**

A motor converts electrical energy into .... The loudspeaker does this by

converting electrical energy into ....

This makes use of the magnetic property of .... which interacts

with the .... causing a varying force. **(4 marks)**

3 Describe how the varying force occurring in a loudspeaker produces sound.

> Think about where the varying force causes movement in the speaker leading to sound being produced in the air.

....

....

....

....

.... **(3 marks)**

4 (a) Describe the result of change in frequency of the vibration of the cone.

....

.... **(1 mark)**

(b) Describe the result of a change in the amplitude of the vibration of the cone.

....

.... **(1 mark)**

Transformers

1 **Two** types of transformers are used in the National Grid. Name them and describe their use.

....

....

....

.... **(2 marks)**

2 Explain how increasing the potential difference before electricity is transmitted across the National Grid helps to reduce wasted energy.

> **Guided**

When potential difference is increased the current is ....

The .... produces less ....

and therefore less ....

.... **(4 marks)**

3 A laptop computer needs a voltage of 19 V. It is connected to the 230 V mains electricity supply using a transformer with 380 turns on the secondary coil. Calculate the number of turns on the primary coil of the transformer. Use the correct equation from the Physics Equations Sheet.

> You will find this equation from the Physics Equation Sheet helpful: $V_p / V_s = n_p / n_s$

....

....

....

Number of turns on primary coil = ... **(3 marks)**

4 A transformer has 600 turns on its primary coil and 20 turns on its secondary coil. The primary voltage is 360 V.

(a) Explain what type of transformer this is.

....

.... **(2 marks)**

(b) Calculate the secondary voltage induced in the secondary coil.

....

....

....

Secondary voltage = ... V **(3 marks)**

Extended response – Magnetism and electromagnetism

Describe how an experiment could show the effect and strength of a magnetic field around a long straight conductor and what would be observed when the circuit was connected.

> You will be more successful in extended response questions if you plan your answer before you start writing.
>
> The question asks you to give a detailed explanation of the magnetic field generated by a long straight conductor. Think about:
>
> - How you would safely connect the conductor to enable circuit measurements to be taken.
> - The methods you could use to determine the direction of a magnetic field.
> - The shape of the magnetic field that you would expect to find.
> - How you would interpret the field patterns of long straight conductors.
> - The variable that would influence the strength of the magnetic field around a long straight conductor.
> - How the influence of the magnetic field of a long straight conductor changes.
>
> You should try to use the information given in the question.

....

....

....

....

....

....

....

....

....

....

....

....

....

....

....

....

....

.... **(6 marks)**

The Solar System

1 Name the main components of the Solar System.

> **Guided**

S............................	P............................	d............................	m............................

(1 mark)

2 Explain the difference between a planet and a moon.

....

.... **(2 marks)**

3 Pluto was discovered in 1930 by Clyde W. Tombaugh and it was classified as a planet. Pluto has now been re-classified. Explain the re-classification and why it was necessary.

> Consider the impact of recently improved observation techniques.

....

....

....

.... **(3 marks)**

4 (a) Describe a difference between the Milky Way and the Solar System in terms of stars.

> **Guided**

The Solar System has only ....

but the Milky Way contains .... **(2 marks)**

(b) Describe another difference between the Solar System and the Milky Way.

The Solar System is a simple ....

but the Milky Way is a .... **(2 marks)**

5 (a) Name the **two** types of planets that exist in the Solar System.

1. ....

2. .... **(2 marks)**

(b) Identify the planet in the list below that is a different type from the other three. Tick **one** box.

☐ Mercury   ☐ Saturn

☐ Mars   ☐ Venus   **(1 mark)**

(c) Give a reason for your answer to (b).

.... **(1 mark)**

(d) Describe the difference between the Earth and Neptune in terms of their surface and atmosphere.

....

.... **(2 marks)**

The life cycle of stars

1   Name the process that occurs in a star when hydrogen nuclei combine to form helium nuclei.

.... **(1 mark)**

2   Identify the correct stage in the lifecycle of the Sun before it became a main sequence star.

☐   Black dwarf

☐   Protostar

☐   Red giant

☐   White dwarf   **(1 mark)**

3   Compare the evolution of stars with masses similar to the Sun to the evolution of stars with more than four times the mass of the Sun.

> Start at main sequence to describe the stages of → giant → collapse → eventual fate.

....

....

....

....

....

....

....

....

.... **(5 marks)**

4   Describe how all elements in the periodic table in the Universe are linked to star processes.

> Consider how fusion processes have led to the formation of elements and how they have been distributed throughout the Universe.

Guided

All of the naturally occurring elements in the periodic table are produced by ....

....

Elements heavier than .... are produced in ....

These explosions of .... then distribute the

elements .... **(5 marks)**

Satellites and orbits

1 Moons that orbit planets are also known by which other name? Tick **one** box.

☐ Artificial satellites

☐ Asteroids

☐ Dwarf planets

☐ Natural satellites   **(1 mark)**

2 Explain the difference between the orbit of a planet and the orbit of a moon.

> Our Solar System is an example of a star that has its own planets, most with their own moons.

....

....

....

.... **(2 marks)**

3 (a) Suggest a similarity in the orbit of planets, moons and satellites.

Planets, moons and artificial satellites all move in .... orbits.   **(1 mark)**

(b) Suggest a difference between the orbit of the Moon and that of an artificial satellite.

A moon orbits at a .... from its planet but the orbit of an

artificial satellite can be changed by adjusting its ....

and the .... of its orbit.   **(3 marks)**

4 (a) Describe how the stable orbit of a satellite around the Earth may be achieved.

....

....

....

.... **(3 marks)**

(b) Describe why a satellite in a stable orbit will move with changing velocity but unchanged speed.

....

....

....

.... **(2 marks)**

(c) Describe how a stable orbit may be achieved if the speed of a satellite changes.

....

.... **(3 marks)**

Red-shift

1 Identify the observation made when light is red-shifted. Tick **one** box.

☐ The light waves travel faster.

☐ The frequency increases.

☐ Light is expanding.

☐ The wavelength increases. **(1 mark)**

2 The diagram shows light from the Sun and light from a distant galaxy. Describe how evidence from a galaxy, observed to have a red-shifted spectrum, gives evidence that the Universe is expanding.

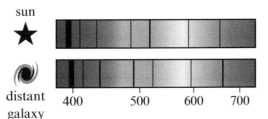

A galaxy with a red-shifted spectrum indicates that the galaxy is ....

.... The further the black lines are shifted ....

.... As most galaxies are red-shifted this would suggest

.... **(3 marks)**

3 The observation of red-shift has enabled scientists to build and develop models to try to understand the evolution, scale and structure of the Universe.

(a) Name the model that scientists have developed, from their observations of red-shift, to explain the origin of the Universe.

.... **(1 mark)**

(b) Distant galaxies are thought to be moving away (receding) faster than the closer galaxies. Describe what type of observation led scientists to this conclusion.

....

.... **(2 marks)**

(c) Describe what observations of the expanding Universe might suggest about its origin.

....

.... **(2 marks)**

(d) (i) Give the names of what appear to bend light and hold galaxies together but cannot be detected by electromagnetic radiation.

....

.... **(2 marks)**

(ii) Describe what may be happening to the expansion of the Universe and why it is so difficult to explain.

....

.... **(2 marks)**

Extended response –
Space physics

 The Big Bang theory is currently widely accepted as the best explanation of the origin of the Universe. Explain why scientists currently accept the Big Bang theory and why it is referred to as the 'current best theory'.

> You will be more successful in extended response questions if you plan your answer before you start writing.
>
> The question asks you to give a detailed explanation of why scientists believe that the Big Bang theory provides the best explanation for the origins of the Universe. Think about:
>
> - A description of what the Universe was believed to have started from.
> - What has happened since the start of the Universe.
> - The evidence that red shift supplies in support of the Big Bang theory.
> - A conclusion that can be reached as a result of red-shift observations.
> - Why scientists have been able to accept the Big Bang theory.
> - Observations that may cast doubt on the Big Bang theory.
>
> You should try to use the information given in the question.

....

....

....

....

....

....

....

....

....

....

....

....

....

....

....

....

....

....

.... **(6 marks)**

Timed Test 1

Time allowed: 1 hour 45 minutes

Total marks: 100

AQA publishes official Sample Assessment Material on its website. The Timed Test has been written to help you practise what you have learned and may not be representative of a real exam paper.

1 (a) A scientist works in the countryside to investigate levels of radioactivity on a remote moor.

He uses a Geiger–Müller detector to measure the count rate.

The detector points across the countryside and the level recorded is an average of 25 counts per minute.

   (i) Name the radiation that the scientist is measuring.   **(1 mark)**

   (ii) Give **two** sources of this radiation.   **(2 marks)**

   (iii) Suggest why the scientist takes a number of readings and then calculates an average.   **(1 mark)**

(b) The scientist conducts a second experiment in the laboratory. First, he measures the background radiation only. Then he tests a sample of radioactive material. He removes the readings for background radiation from those taken from the radioactive sample and records the corrected count rate in a table.

Time in minutes	Corrected count rate in counts / min
0	1030
1	760
2	515
3	387
4	258
5	194
6	129
7	98
8	65

   (i) Plot a graph of the corrected count rate against time.   **(3 marks)**

   (ii) Add a curve of best fit.   **(1 mark)**

   (iii) Determine the half-life of the radioactive source.   **(1 mark)**

(c) Describe the nature of alpha, beta and gamma radiation in terms of ionising properties.   **(3 marks)**

2 A lift travels between floors in a building.

(a) The lift moves from the ground floor to the fourth floor through a height of 15 m in 20 s. The mass of the lift is 750 kg.

   (i) Calculate how much energy the lift gains in moving from the ground floor to the fourth floor. Take g to be 10 N/kg.   **(2 marks)**

(ii) Calculate the average power that the elevator exerts in moving the mass of 750 kg to the fourth floor. Choose the correct unit from the box.

| watts / W | newtons / N | joules / J |

(3 marks)

(b) As the lift moves upwards, not all of the energy supplied is usefully transferred. Suggest where some of the wasted energy is transferred to. **(1 mark)**

(c) Calculate the kinetic energy of the elevator in moving from the ground floor to the fourth floor. **(4 marks)**

(d) Explain why this mechanical process could be described as wasteful. **(1 mark)**

(e) Give an example of wasted energy in the process described in (d). **(1 mark)**

3 (a) A kettle is rated at 2000 W and is designed to operate on a 230 V mains supply.

  (i) Write the equation that links power, voltage and current. **(1 mark)**

  (ii) Calculate the current drawn by the kettle. **(2 marks)**

  (iii) Describe what could happen to the kettle if a fuse were to be fitted that has:

    too high a rating **(1 mark)**

    too low a rating. **(1 mark)**

  (iv) Which fuse should be fitted to the kettle to make it safe? Tick **one** box.

  ☐ 1 A   ☐ 3 A   ☐ 10 A   ☐ 13 A   **(1 mark)**

(b) Calculate the charge moved in the kettle in (a) each second. Choose the correct unit from the box.

| C | J | W |

(3 marks)

4 An electrical heater supplies electrical energy to a copper block of mass 2000 g at 12 V with a current of 12 A for 2 minutes.

(a) Calculate the energy supplied to the heater. Use the correct equation from the Physics Equation Sheet. Choose the correct unit from the box.

| joules / J | amps / A | volts / V |

(3 marks)

(b) Calculate the temperature rise for the block of copper when supplied with 2880 J. The specific heat capacity of copper is 385 J/kg °C. Use the correct equation from the Physics Equation Sheet. Choose the correct unit from the box.

| joules / J | degrees Celsius / °C | watts / W |

(3 marks)

(c) Suggest why the actual temperature rise may be lower than the predicted value. **(1 mark)**

(d) (i) Explain how unwanted energy transfer in the heating of a metal block could be reduced. **(1 mark)**

  (ii) Name the type of material that should be used to reduce thermal energy transfer. **(1 mark)**

5 (a) Complete the nuclear equations, below, for alpha and beta decay.

  (i) $_{91}^{.....}\text{Pa} \rightarrow {}^{227}_{.....}\text{Ac} + {}^{4}_{.....}\text{He}$ **(1 mark)**

  (ii) Name the type of decay in (i). **(1 mark)**

  (iii) $_{87}^{.....}\text{Fr} \rightarrow {}^{.....}_{2}\text{He} + {}^{207}_{.....}\text{At}$ **(1 mark)**

  (iv) Name the type of decay in (iii). **(1 mark)**

  (v) $_{11}^{24}\text{Na} \rightarrow {}^{24}_{.....}\text{Mg} + {}^{.....}_{-1}\text{e}$ **(1 mark)**

  (vi) Name the type of decay in (iv). **(1 mark)**

  (vii) $_{.....}^{201}\text{Au} \rightarrow {}^{.....}_{80}\text{Hg} + {}^{0}_{-1}\text{e}$ **(1 mark)**

(viii) Name the type of decay in (vii). **(1 mark)**

(b) (i) Name any changes that occur as a result of alpha nuclear decay in terms of mass and charge. **(1 mark)**

    (ii) Name any changes that occur as a result of beta nuclear decay in terms of mass and charge. **(1 mark)**

6 A stationary fuel tanker delivers fuel to a plane. The fuel line is connected to an earth wire to prevent any static charge building up.

(a) (i) Explain why static electricity may build up in the fuel line. **(2 marks)**

    (ii) Explain why the build-up of static charge could be dangerous in this situation. **(2 marks)**

(b) Give the **two** conditions that result in a build-up of static charge on an object. **(2 marks)**

(c) Discuss how an electric field may be generated and the effect that this may have on a charged object within that field. Include a diagram to illustrate the electric field generated from a positive point source. **(6 marks)**

7 The circuit diagram shows three identical lamps and one cell.

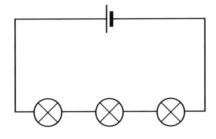

(a) The cell provides 1.5 V. Explain what the potential difference across each lamp is. **(2 marks)**

(b) Add a voltmeter to the circuit diagram to show how you could measure the potential difference across one lamp. **(1 mark)**

(c) Name **one** component that could be added to make the lamps brighter. **(1 mark)**

(d) Suggest a more effective way of connecting the lamps to make them brighter. **(1 mark)**

(e) The circuit diagram below shows a circuit to test a thermistor.

    (i) Describe how resistance changes in a thermistor.
    (3 marks)

    (ii) Give a use for a thermistor in a domestic circuit.
    (1 mark)

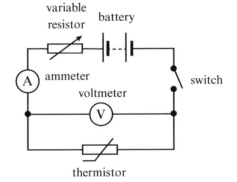

8 (a) Draw diagrams in the boxes to show the differences in particle arrangement for elements of solids, liquids and gases. **(2 marks)**

solid	liquid	gas

(b) Explain the difference between solids, liquids and gases in terms of the kinetic energy of particles.

(6 marks)

9 (a) A student conducts an investigation into density. She measures the dimensions of three rectangular blocks with a ruler to calculate the volume. The first block has a volume of 0.2 m³ and a mass of 1500 g. The second block has a volume of 0.15 m³ and a mass of 0.7 kg. The third block has a volume of 0.2 m³ and a mass of 0.7 kg. Determine which block has the highest density.

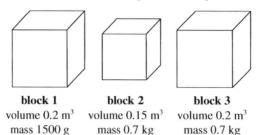

block 1
volume 0.2 m³
mass 1500 g

block 2
volume 0.15 m³
mass 0.7 kg

block 3
volume 0.2 m³
mass 0.7 kg

(4 marks)

(b) The student now investigates the density of three irregular rocks. She fills a Eureka can with water, places the first rock into the can and collects the water displaced in a measuring cylinder.

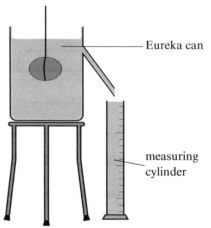

Eureka can

measuring cylinder

(i) Explain what the student must do to make sure that the reading on the measuring cylinder is as accurate as possible. **(2 marks)**

(ii) Describe the other measurement that the student must make to be able to calculate the density of the rock. **(2 marks)**

10 (a) Describe the differences between Thompson's model of the atom (the 'plum pudding' model) and Rutherford's model of the atom (the 'nuclear' model). **(4 marks)**

(b) Describe the alpha-scattering experiment, carried out by Geiger and Marsden, which produced evidence to suggest and support a new model for the atom. **(6 marks)**

Timed Test 2

Time allowed: 1 hour 45 minutes

Total marks: 100

AQA publishes official Sample Assessment Material on its website. The Timed Test has been written to help you practise what you have learned and may not be representative of a real exam paper.

1 Charlie and Ravi plan to set up an experiment to measure speed. They have a trolley, an inclined ramp, a ruler and a stopwatch.

  (a) Describe a method that they could use to measure the speed of the trolley using the apparatus above.

                                                                   (4 marks)

  (b) Suggest other apparatus that the students could use to improve the precision of the data collected.

                                                                   (2 marks)

  (c) Charlie and Ravi then extended their experiment to investigate the effect of the independent variable. The table below shows data collected by the students.

1	2	3	4
in cm	in m	in s	in [unit]
5	1.80	3.2	
10	1.80	2.4	
15	1.80	1.8	
20	1.80	1.4	
25	1.80	1.0	
30	1.80	0.4	
35	1.80	0.2	

    (i)   Suggest what the new independent variable is.   **(1 mark)**

    (ii) Write titles for columns 1–3 in the table.   **(1 mark)**

    (ii) Write a heading in column 4 that would be useful for recording processed data. Give the unit. **(1 mark)**

  (d) After travelling down the ramp, the trolley comes to a stop by itself.

    (i)   Name/describe the forces involved in causing the trolley to stop.   **(2 marks)**

    (ii) Describe how these forces could be reduced.   **(2 marks)**

2 A crash test car of mass 1000 kg is driven at the design-testing centre to examine impact forces. The car starts from rest and accelerates to its final speed.

  (a) Write the equation used to calculate the acceleration of the car towards the crash barrier in a time t.   **(1 mark)**

  (b) The car accelerates uniformly from 0 m/s to 10 m/s over a time of 20 s.

    Calculate how far the car will travel.   **(2 marks)**

  (c) Calculate the momentum of the car as it crashes into the crash barrier. Give the unit.   **(3 marks)**

  (d) Which of the following would decrease the momentum in a collision? Tick **one** box.

  ☐ Decreasing the crumple zones present

  ☐ Increasing the velocity of the vehicle

  ☐ Decreasing the mass of the vehicle

  ☐ Decreasing the time taken to change momentum   **(1 mark)**

3 (a) Which stage is **not** part of the life cycle of a star of similar mass to the Sun?

☐ Main sequence

☐ Nebula

☐ Supernova

☐ White dwarf   **(1 mark)**

(b) (i) Define the term fusion.   **(1 mark)**

(ii) Describe the **two** outcomes of fusion processes in stars relating to naturally occurring elements.   **(2 marks)**

(iii) Describe how supernovae play a role in the production and distribution of elements.   **(2 marks)**

4 The diagram below shows the graph of a radio wave.

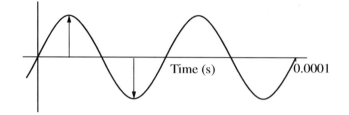

(a) (i) Give the equation linking time period and frequency.   **(1 mark)**

(ii) Identify the time period of the wave by adding 'T' with arrows to indicate one complete time period.   **(1 mark)**

(iii) Calculate the frequency of the wave shown in the diagram.   **(3 marks)**

(b) Fishing trawlers use sound waves to locate shoals of fish.

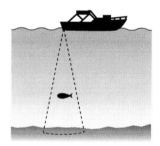

(i) Give the name of the process of using sound in water to determine the position of an object relative to a transmitter.   **(1 mark)**

(ii) Explain how the ship can use sound waves to locate the shoal of fish.   **(3 marks)**

5 (a) Give the property of light that enables it to be used in fibre optic communications.   **(1 mark)**

(b) The diagram shows a light wave being refracted through a glass block.

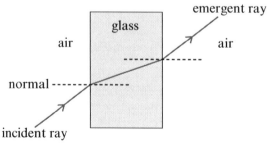

(i) Label the angle of incidence (i) and the angle of refraction (r) on the diagram.   **(2 marks)**

(ii) Explain why the wave changes direction.   **(2 marks)**

(c) Light entering a convex lens is shown in the diagram. Complete the diagram to show the location and orientation of the image. **(3 marks)**

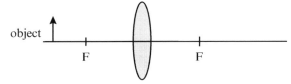

6 Edwin Hubble discovered that light from galaxies that was seen from Earth was red-shifted. He found this by looking at the absorption spectra of stars and galaxies.

(a) Describe how the characteristics of a light source that is red-shifted is different from light that is not red-shifted. **(2 marks)**

(b) Suggest the relationship between the recessional speed of galaxies in relation to the observed wavelength of their light. **(2 marks)**

(c) Explain how the observations of red shift provide evidence for the Big Bang model of the universe. **(2 marks)**

(d) Red shift alone does not fully explain the Big Bang model of the universe. Describe the other piece of evidence that supports this model. **(2 marks)**

(e) Astronomers think that galaxies should rotate much faster than they do. Describe the ideas that scientists have proposed that may account for this and why they cannot be explained. **(6 marks)**

7 The Solar System is a collection of many bodies that orbit round other bodies in space.

(a) Explain the difference between the orbits of planets, including dwarf planets, and the orbits of moons. **(2 marks)**

(b) Most planets of the Solar System have natural satellites. Compare Jupiter with Mars in terms of relative planetary size and relative number of natural satellites. **(2 marks)**

(c) Explain why a satellite can have a constant speed but a changing velocity. **(2 marks)**

(d) Explain the life cycle of a star like our Sun, which is the centre of our Solar System. **(6 marks)**

8 The diagram shows a wire passing through a card. A compass is placed in different spots around the wire to determine the direction of the magnetic field.

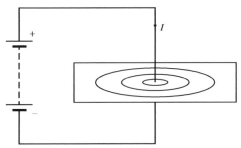

(a) (i) Draw arrows on the diagram to show the magnetic field produced when the current flows through the coil. **(2 marks)**

(ii) Name the rule that can be used to determine the direction of the field. **(1 mark)**

(b) (i) The wire is now turned into a solenoid. Draw the new magnetic field that is produced. **(4 marks)**

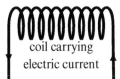

coil carrying electric current

(ii) Suggest a use for this device. **(1 mark)**

(c) A 0.75-m length of wire carrying a current of 3 A, is placed between two magnets at right angles to the field of 0.5 T. Calculate the force experienced by the wire. Give the unit. Use the correct equation from the Physics Equation Sheet. **(3 marks)**

9 The diagram shows a step-up transformer.

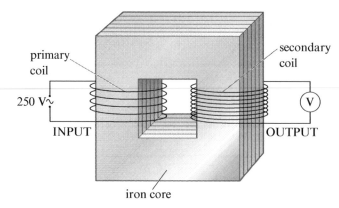

(a) Explain how step-up transformers improve the efficiency of energy transfers through the National Grid. **(2 marks)**

(b) When $n_p = 150$ and $n_s = 4500$, calculate the induced potential difference in the secondary coil when the primary potential difference is 250 V. Use the correct equation from the Physics Equation Sheet. **(3 marks)**

(c) Explain what happens to the electricity supply before the electricity can enter a domestic building from the National Grid. **(2 marks)**

(d) (i) $V_p = 4600$ V, $I_p = 5$ A and $V_s = 230$ V. Calculate the current in the secondary coil of a step-down transformer. Use the correct equation from the Physics Equation Sheet. **(3 marks)**

    (ii) Give the assumption made when calculating the power of a transformer. **(1 mark)**

10 Two students carry out an experiment to investigate the extension of a spring. They add five masses to a spring: one 10 g mass, one 50 g mass and three 100 g masses. They then measure the final extension. When the spring is unloaded, the students find that the spring has stretched.

(a) Identify a problem with investigating inelastic deformation when using this method. **(1 mark)**

(b) Suggest **two** ways of improving the experiment. **(2 marks)**

(c) The table below shows results from a different experiment for the loading of a 32-mm spring. Complete the missing values in the table.

Weight in N	Extension in mm
0	32
0.1	
	40
0.3	
0.4	48
	52
0.6	56

(2 marks)

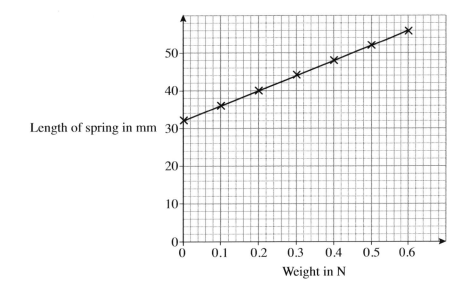

Length of spring in mm

Weight in N

(d) Explain how the graph illustrates the relationship between the length of the spring and the weight added.
You should refer to the graph in your answer.   **(4 marks)**

Answers

Extended response questions

Answers to 6-mark questions are indicated with a star (*).

In your exam, your answers to 6-mark questions will be marked on how well you present and organise your response, not just on the scientific content. Your responses should contain most or all of the points given in the answers below, but you should also make sure that you show how the points link to each other, and structure your response in a clear and logical way.

1 Energy stores and systems

1 A – gravitational, C – chemical, D – kinetic **(1)** (all three needed for mark)

2 (a) A closed system is an isolated system where no energy flows in or out of the system. **(1)**

   (b) The total energy in a closed system is the same after the transfer as it was before the transfer **(1)** because energy cannot be created or destroyed **(1)**.

3 (a) Store of chemical energy **(1)**.

   (b) Energy transfer by electrical current **(1)**.

   (c) Energy transfer to the surroundings by sound waves and by heating **(1)** (both needed for mark).

4 The total energy available initially in this closed system, in the gravitational potential store, is 250 J **(1)**. As the basket reaches the ground, the gravitational potential store will become 0 J **(1)**, because it has been transferred to a total of 250 J of useful kinetic energy **(1)** and wasted thermal energy **(1)**.

2 Changes in energy

1 D **(1)**

2 Kinetic energy $E_k = \frac{1}{2} m v^2$ so $E_k = \frac{1}{2} \times 70 \times 6^2$ **(1)** = 1260 **(1)** J / joules **(1)**

3 Convert units: 15 cm = 0.15 m **(1)**

   Energy transferred $E = \frac{1}{2} k x^2 = \frac{1}{2} \times 200$ N/m $\times (0.15$ m$)^2$ **(1)** = 2.25 **(1)** J

4 Change weight to mass first, using $m = F / g$ so 600 / 10 = 60 kg **(1)**

   Gravitational potential energy (GPE) = 60 kg $\times$ 10 N/kg $\times$ 0.7 m **(1)** = 420 **(1)** J

3 Energy changes in systems

1 Specific heat capacity = change in thermal energy $\div$ (mass $\times$ change in temperature) or $(c = \Delta E \div (m \times \Delta T)$ **(1)**

2 $\Delta E = m c \Delta\theta$ **(1)** = 0.8 kg $\times$ 4200 J / kg °C $\times$ 50 °C **(1)** = 168 000 **(1)** J

3 $\Delta\theta = \Delta E / (m c)$ so $\Delta\theta$ = 20 000 J / (1.2 kg $\times$ 385 J/kg °C) **(1)** = 43 **(1)** so

   Change in temperature of the copper = 43 °C **(1)**

4 $E_{in} = P t$ = 30 W $\times$ 540 s **(1)** = 16 200 J **(1)**

   $\Delta E_{in} = \Delta E_{out}$ so

   Specific heat capacity c = 16 200 J / (0.8 kg $\times$ 25 °C) **(1)** = 16 200/20 = 810 **(1)** J / kg °C

4 Specific heat capacity

1 (a) the amount of energy required to raise the temperature of 1 kg of material by 1 K (or 1 °C) **(1)**

   (b) Energy supplied, mass and change in temperature. **(1)**

2 (a) Place a beaker on a balance, zero the balance and add a measured mass of water **(1)**. Take a start reading of the temperature **(1)**. Place the electrical heater into the water and switch on **(1)**. Take a temperature reading every 30 seconds **(1)** until the water reaches the required temperature **(1)**.

   (b) Measure the current supplied, the potential difference across the heater and the time for which the current is switched on **(1)**. Use these values to calculate the thermal energy supplied using the equation $E = V \times I \times t$ **(1)**.

   (c) Add insulation around the beaker **(1)** so less thermal energy is transferred to the surroundings and a more accurate value for the specific heat capacity of the water may be obtained. **(1)**

3 Plot a graph of temperature against time **(1)**. The changes of state are shown when the graph is horizontal (the temperature is not increasing) **(1)**.

5 Power

1 A **(1)**

2 Energy transferred = 15 000 J, time taken = 20 s

   $P = E / t$

   so power P = 15 000 J / 20 s **(1)** = 750 **(1)** W

3 (a) $E_g = m g h$ so E_g= 60 $\times$ 10 $\times$ (0.08 $\times$ 20) = 960 **(1)** joules / J **(1)**

   (b) $P = E / t$ so P = 960 / 12 = 80 watts / W **(1)** (allow value for energy calculated in (a))

4 (a) For 3 W motor: t = 360 J $\div$ 3 W **(1)** = 180 **(1)** s.

   (b) For 5 W motor: t = 360 J $\div$ 5 W **(1)** = 72 **(1)** s

6 Energy transfers and efficiency

1 (a) Concrete **(1)**

   (b) Low relative thermal conductivity means that a material will have a slow **(1)** rate of transfer of thermal energy. **(1)**

2 (a) Thicker walls provide more material for the thermal energy **(1)** to travel through from the inside to outside, so the rate of thermal energy **(1)** loss is less, keeping the houses warmer.

   (b) Thicker walls provide more material for the thermal energy **(1)** to travel through from the outside to inside, so the rate of thermal energy transfer **(1)** is less, keeping the house cool.

3 (a) The useful energy transferred to the box = 100 J; total energy used by the motor = 400 J; efficiency = 100 J / 400 J **(1)** = 0.25. **(1)** (accept $\times$ 100 = 25%.)

   (b) Using lubrication between moving surfaces will reduce friction **(1)** and therefore reduce wasted thermal energy **(1)**.

4 A **(1)**

7 Thermal insulation

1 (a) Any **five** of the following points: select a minimum of four beakers, one to remain unwrapped as a control **(1)**, wrap three or more beakers with the same mass / thickness of insulating material **(1)**, provide insulating bases and lids (with a hole for the thermometer) **(1)**, add the same amount of boiling or very hot water to the beakers **(1)**, place thermometers in each beaker and record the temperature of the water at the highest point **(1)**, start stop watch **(1)**, and if recording a temperature curve, take temperature reading every minute (or other time increment) **(1)**.

   (b) Independent variable – type of insulation selected **(1)**; dependent variable – temperature **(1)**

   (c) Control variables: any four from – mass / thickness of insulating material, volume of water, starting temperature of the water, size of beaker, material of beaker, time of experiment. (For four **(2)**, for three **(1)**)

2 Any one of the following: hot water in eyes can cause damage – always wear eye goggles **(1)**; hot water can cause scalds – place kettle close to beakers **(1)**; control beaker, and others, can cause burns – do not touch **(1)**; spilt water can cause slippage – report and wipe up **(1)**; and glass thermometers can be broken and cause cuts – handle with care (same if using glass beakers) **(1)**.

3 A digital thermometer can improve accuracy in reading the temperature. **(1)**

4 The greater the thickness of insulating material, the slower the rate at which the hot water cools **(1)**. The lower the thermal conductivity of the insulating material, the lower the rate at which the hot water will cool **(1)**.

8 Energy resources

1 (a) A hydroelectric power station is a reliable producer of electricity because it uses the gravitational potential energy of water which can be stored until it is needed **(1)**. As long as there is no prolonged drought / lack of rain the supply should be constant **(1)**.

   (b) Any **one** of the following: hydroelectric power stations have to be built in mountainous areas / high up (compared with supply areas, so that the gravitational potential energy can be captured) **(1)**; the UK has very few mountainous areas like this **(1)**; and limited to areas such as North Wales and Scottish Highlands **(1)**.

2 (a) When carbon dioxide is released into the atmosphere it contributes to the greenhouse effect / build-up of CO_2 **(1)** which is believed to contribute to global warming **(1)**.

   (b) Sulfur dioxide and nitrogen oxides have been found to dissolve in the water droplets in rainclouds, increasing their acidity **(1)**; this can kill plants and forests / lakes or dissolve the surfaces of historical limestone buildings **(1)**.

   (c) Coal mines / oil / gas wells create environmental scars on the landscape **(1)**. Vehicles used to transport fossil fuels add to environmental pollution **(1)**. Alternative answers may include: Accidents in the extraction of oil from deep sea reserves can result in sea pollution **(1)**. New methods of extraction may impact on previously unused areas of the environment e.g. fracking **(1)**. Large transport networks may be needed to transport fuels **(1)**.

9 Patterns of energy use

1 (a)
   1. After 1900, the world's energy demand rose / increased **(1)** as the population grew.
   2. There was development in industry / demand in energy supply. **(1)**
   3. The rise of power stations using fossil fuels added to demand. **(1)**

   (b) (i) Coal, oil and natural gas **(1)** (all three needed)
       (ii) Any two from: population has increased so domestic energy use has increased **(1)**; industry has grown, requiring more energy **(1)**; vehicle use and travel have grown, requiring more energy **(1)** (and any other valid reason)
       (iii) Nuclear research only began from the 1940s onwards. **(1)**
       (iv) Hydroelectric. **(1)**

2 As the population continues to rise the demand for energy will also continue to rise **(1)**. Current trends show that the use of fossil fuels is the major contributor to the world's energy resources **(1)**. These are running out and no other energy resource has, so far, taken their place **(1)**. This could lead to a large gap between demand and supply **(1)**. (Any other valid reason)

10 Extended response – Energy

The answer should include some of the following points: **(6)**
- Refer to the change in gravitational potential energy (E_p) as the swing seat is pulled back/ raised higher.
- Before release, the E_p is at maximum/kinetic energy (E_k) of the swing is at a minimum.
- When the swing is released, the E_p store falls and the E_k store increases.
- E_k is at a maximum at the mid-point, E_p is at a minimum.

- The system is not 100% efficient; some energy is dissipated to the environment.
- Friction due to air resistance and/or at the pivot results in the transfer of thermal energy to the surroundings/environment.
- Damping, due to friction, will result in the E_k being transferred to the thermal energy store of the swing and hence to the environment.
- Eventually all the E_p will have been dissipated to the surroundings/environment (so is no longer useful).

11 Circuit symbols

1 C **(1)**

2 (a) 1. Resistor **(1)**
       2. Fuse **(1)**
       3. Variable resistor **(1)**
   (b) Fuse **(1)**

3

Component	Symbol	Purpose
ammeter	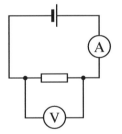	measures electric current **(1)**
fixed resistor **(1)**		provides a fixed resistance to the flow of current
diode **(1)**		allows the current to flow one way only
switch **(1)**	or	allows the current to be switched on / off

(Each correctly completed row gains 1 mark.)

4 Diagram showing series circuit diagram with battery / power supply **(1)**. Resistor **(1)** with ammeter in series **(1)**. Voltmeter connected in parallel across the resistor **(1)**.

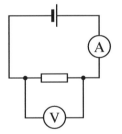

12 Electrical charge and current

1 (a) An electric current is the rate **(1)** of flow of charge (electrons in a metal) **(1)**.
   (b) Charge $Q = I\,t = 4\text{ A} \times 8\text{ s}$ **(1)** = 32 **(1)** coulombs / C **(1)**

2 (a) (i) The current is the same in all parts of a series circuit so the readings on ammeter 1 and ammeter 3 will be the same as that shown for ammeter 2. **(1)**
       (ii) Add another cell / increase the energy supplied. **(1)**
   (b) Cell **(1)**

3 (a) Any series circuit diagram with a component (e.g. lamp) **(1)** and an ammeter **(1)**.
   (b) stopwatch / timer **(1)**

13 Current, resistance and pd

1 D **(1)**

2 Ohm's law states: The current flowing through a resistor **(1)** at constant temperature is directly proportional to the potential difference across the resistor. **(1)**

3 (a) Resistance $R = V / I = 12\text{ V} / 0.20\text{ A}$ **(1)** = 60 Ω **(1)**
   (b) Current $I = V / R$ so $I = 22\text{ V} / 55\ \Omega$ **(1)** = 0.40 A **(1)**

4 (a) Line A – straight line through origin **(1)**
       Line B – straight line through origin – different gradient **(1)**
   (b) Line with smaller gradient **(1)**

14 Investigating resistance

1 B **(1)**

2 $R = V / I$ so $R = 90 / 1.5$ **(1)** = 60 **(1)** Ω.

3 All circuit symbols correct (2 cells, 2 lamps, 1 ammeter, 1 voltmeter, wire) **(1)**. Ammeter connected in series **(1)**. Voltmeter connected in parallel with one lamp **(1)**.

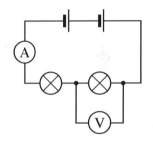

4 A Fixed resistor: The temperature remains constant so the resistance remains constant, as shown by the straight line on the graph. **(1)** (both needed)
   B Filament lamp: As potential difference increases, the filament gets hot, so resistance increases, as shown by the curved line on the graph. **(1)** (both needed)
   C Diode: The current flows in only one direction and the resistance is constant, as shown by the straight line on the graph. **(1)** (both needed)

15 Resistors

1 (a) As the potential difference increases, the current increases **(1)** in a linear / proportional relationship **(1)**.
   (b) As the potential difference increases the current increases **(1)** but the gradient of the line gets less steep / shallower, or increase in current becomes smaller as the potential difference continues to increase **(1)**.

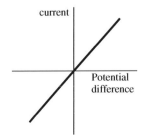

2   (a) Fixed resistor: as graph A in Q1 **(1)**

Filament lamp: as graph B in Q1 **(1)**

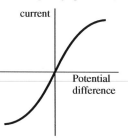

(b) The different shaped graphs are because the fixed resistor is ohmic / obeys Ohm's law, so the current and potential difference have a proportional relationship **(1)**; the filament lamp does not obey Ohm's law so the relationship between current and potential difference is not proportional / current begins to level off as potential difference increases **(1)**.

3   Data can be collected using an ammeter to measure current **(1)** and a voltmeter to measure potential difference **(1)**. A wire should be included and a fixed resistor **(1)** to prevent overheating. A range of potential difference **(1)** measurements should be made so that resistance can be calculated using the equation $R = V / I$ **(1)**.

16 LDRs and thermistors

1

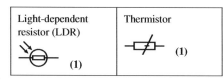

Light-dependent resistor (LDR)	Thermistor
(1)	**(1)**

2   (a) The resistance goes down as the light becomes more intense (brighter) (more current flows). **(1)** (b) The resistance goes down as the temperature goes up (more current flows). **(1)**

3   The thermistor reacts to rise in temperature **(1)** in the engine. Above a certain temperature, it allows current **(1)** in the circuit to flow to a fan, which cools **(1)** the engine.

4   In the light, the resistance of the LDR is low **(1)** so less current flows through the lamp, turning the light off (in the light, more current will flow through the LDR instead of through the lamp) **(1)**.

17 Investigating *I–V* characteristics

1   (a) Ammeter connected in series. **(1)** Voltmeter connected in parallel across the component to be tested. **(1)**

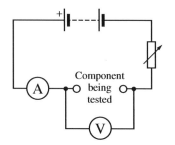

(b) (i)   Potential difference (*V*) **(1)**

(ii) Current (*I*) **(1)** (can be in reverse order)

(c) The terminal connections should be reverse to obtain negative values. **(1)**

(d) *y*-axis: current (*I*), *x*- axis: potential difference (*V*) **(1)** (both needed)

2   (a) wire / ohmic resistor **(1)**

(b) filament lamp **(1)**

(c) diode **(1)**

3   Ohm's law: resistance = potential difference / current **(1)**

R (Ω or ohms) = V (V or volts) / I (A or amps/ amperes) (1 mark for all three symbols, 1 mark for all three units)

4   Resistors can become hot and cause burns or fire. **(1)**

18 Series and parallel circuits

1   (a) In a series circuit the current flowing through each component is the same **(1)**. In a parallel circuit, the current is shared between the components **(1)**.

(b) Series: $A_2 = 3$ A, $A_3 = 3$ A **(1)** (both needed)

Parallel: $A_2 = 1$ A, $A_3 = 1$ A, $A_4 = 1$ A **(1)** (all three needed)

2   (a) In a series circuit, the total potential difference supplied is shared between the components **(1)**. In a parallel circuit, the potential difference across each component is the same as the potential difference supplied **(1)**.

(b) Series: $V_2 = 3$ V, $V_3 = 3$ V, $V_4 = 3$ V **(1)** (all three needed)

Parallel: $V_2 = 9$ V, $V_3 = 9$ V, $V_4 = 9$ V **(1)** (all three needed)

3   The total resistance of two or more resistors arranged in series is equal to **(1)** the sum of the resistance of each component **(1)**. The total resistance of two or more resistors arranged in parallel is less than **(1)** the resistance of the smallest individual resistor **(1)**.

19 ac and dc

1   (a) Direct potential difference is constant **(1)** and the current flows in the same direction **(1)**.

(b) Alternating potential difference is changeable **(1)** and the current constantly changes direction **(1)**.

2   (a) $P = E / t$ so $E = P t = 2000 \times (15 \times 60)$ s **(1)** = 1 800 000 J **(1)**

(b) $E = 10 \times (6 \times 60 \times 60)$ s **(1)** = 216 000 J **(1)**

3   There should be one horizontal line anywhere on the screen. **(1)**

4   (a) Toaster: $P = E / t = 120\,000$ J / 60 s **(1)** = 2000 W **(1)**

(b) Kettle: $P = 252\,000$ J / 120 s **(1)** = 2100 W **(1)**

Therefore the kettle has the higher power rating. **(1)**

20 Mains electricity

1   Earth wire (green and yellow), **(1)** live wire (brown), **(1)** neutral wire (blue), **(1)** fuse **(1)**

2   (a) Mains electricity is delivered through an alternating **(1)** current.

(b) The potential difference between the live wire and neutral wire is about 230 **(1)** V. The neutral wire is at, or close to, 0 **(1)** V.

(c) The earth wire is at 0 V **(1)** and only carries a current if there is a fault. **(1)**

(d) In the UK, the domestic electricity supply has a frequency of 50 **(1)** Hz.

3   When a large current enters the live wire **(1)**, this produces thermal energy **(1)**, which melts the wire in the fuse **(1)** and the circuit is then broken **(1)**.

4   The earth wire is connected to the metal casing **(1)**. If the live wire becomes loose and touches anything metallic, the user is protected because the current passes out through the earth wire **(1)** rather than through the user **(1)**.

21 Electrical power

1   $P = I\,V = 5$ A $\times$ 230 V **(1)**

Power = 1150 **(1)** W

2   (a) $P = I\,V$ so $I = P / V$ **(1)** = 3 W / 6 V **(1)**

Current = 0.5 **(1)** A

(b) $P = I^2 R$ **(1)** = $(0.5$ A$)^2 \times 240$ Ω **(1)**

Power = 60 **(1)** W

3   D **(1)**

4   Find current: $I = V / R$ so $I = 80$ V / 8 Ω = 10 A **(1)**

use either: power $P = I\,V$ so $P = 10$ A $\times 80$ V **(1)** = 800 **(1)** W

or $P = I^2 R$ so $P = 100$ A $\times 8$ Ω **(1)** = 800 **(1)** W

22 Electrical energy

1   $E = Q\,V$ **(1)** = 30 C $\times$ 9 V **(1)** = 270 J **(1)**

2   (a) $Q = I\,t$ so $Q = 0.2$ A $\times 4$ V = 0.8, so charge flow = 0.8 **(1)** C.

(b) $E = Q\,V = 0.8$ C $\times 30$ s **(1)** = 24 so energy transferred = 24 **(1)** joules / J **(1)**

3   (a) The power of a circuit device is a measure of the rate of energy transfer / how fast the energy is transferred by the device. **(1)**

(b) Energy transferred is the amount of power over a given time **(1)** and the power of a device is the product of the current passing through it and the potential difference across it **(1)**.

4   (a) $Q = E / V = 600$ J / 20 V **(1)** = 30 C **(1)**

(b) $Q = I\,t$, so $t = Q / I = 30$ C / 0.15 A **(1)** = 200 **(1)** s

23 The National Grid

1   (a) As the voltage is increased so the current goes down **(1)**, so this reduces the heating effect due to resistance **(1)** and means that less energy is wasted in transmission **(1)**.

(b) Wasting less energy in transmission means that more energy is transferred to where it is needed **(1)**, making the National Grid an efficient way to transmit energy **(1)**.

(c) The voltages are high enough to kill you if you touch or come into contact with a transmission line. **(1)**

2   $P = I\,V = 20\,000\ \text{A} \times 25\,000\ \text{V}$ **(1)** $=$ $500\,000\ \text{kW}$ **(1)**

3   Any two from: step-up transformers increase voltage and so lower the current, reducing the heating losses **(1)**. Wires are thermally insulated **(1)**. Wires of low resistance are used **(1)**.

4   Step-up transformers are used to increase the voltage **(1)** as it leaves the power station for transmission through the National Grid **(1)**. Near homes, step-down transformers are used to reduce the voltage **(1)** to make it safer for use in houses **(1)**.

24   Static electricity

1   The student transfers a charge on to the balloon by transferring electrons from the jumper to the balloon **(1)**. The negative charges on the balloon repel the electrons in the wall **(1)**, inducing a positive charge on the wall **(1)** which attracts the negatively charged balloon **(1)**.

2   B **(1)**

3   (a) Friction transfers electrons to another material. **(1)**

    (b) Friction transfers electrons from another material. **(1)**

4   Insulators do not allow electrons to flow **(1)**. Instead the electrons either collect on the insulator (building up a charge) **(1)** or are knocked off (leaving a positive charge) **(1)**. Conductors allow electrons to flow **(1)**. Mutually repelling electrons will then flow away and this dissipates any charge build-up **(1)**.

25   Electric fields

1   An electric field is a region in space **(1)** where a charged particle may experience a force. **(1)**

2   High density lines of flux **(1)**, radial field **(1)**, arrows pointing outwards **(1)**.

radial field

3   The electric field is strongest close to the charged object **(1)**. The further away from the charged object, the weaker the field **(1)**.

4   (a) (i)   outwards **(1)**

      (ii)   outwards **(1)**

      (iii)   inwards **(1)**

    (b) In an electric field, electrically charged particles experience a force **(1)** that accelerates the particles **(1)**.

5   The student is right **(1)** because the electrically charged insulator will either be negative or positive **(1)** (both needed) due to gaining or losing electrons **(1)**. The charged insulator then becomes a point source creating an electric field **(1)**.

26   Extended response – Electricity

The answer should include some of the following points: **(6)**

- The thermistor can be connected in series with an ammeter to measure current with a

voltmeter connected in parallel across it to measure potential difference.
- Ohm's law can be referred to in calculating the resistance.
- When the temperature is low the resistance of the thermistor will be high, allowing only a small current to flow.
- When the temperature is high the resistance of the thermistor will be low, allowing a larger current to flow.
- The light-dependent resistor can be connected in series with an ammeter to measure current with a voltmeter connected in parallel across it to measure potential difference.
- When light levels are low (dark) the resistance of the light-dependent resistor will be high, allowing only a small current to flow.
- When light levels are high (bright) the resistance of the light-dependent resistor will be low, allowing a larger current to flow.
- Thermistors can be used in fire alarms as a temperature sensor to switch on an alarm.
- Light-dependent resistors can be used in security systems as a light sensor to switch on a light.

27   Density

1   (a) 1.   Solid

      2.   Liquid

      3.   Gas **(1)**

    (b) In a solid, mass per unit volume is higher than for a liquid or a gas because the particles are very close together **(1)**. In a liquid, mass per unit volume is lower than that for a solid, because the particles are further apart, but higher than that for a gas because the particles are closer together **(1)**. In a gas, mass per unit volume is low because the particles are furthest apart **(1)**.

    (c) 1.   In a solid, high mass / number of particles per unit volume means that density is high **(1)**.

      2.   In a liquid, lower mass / number of particles per unit volume (than in a solid) means that density is lower (than in a solid) **(1)**.

      3.   In a gas, very low mass / number of particles per unit volume means that density is very low **(1)**.

2   B **(1)**

3   Volume $= 10\ \text{cm} \times 25\ \text{cm} \times 15\ \text{cm} = 3750\ \text{cm}^3$ **(1)**

  $\rho = m \div V$

  so $m = \rho\,V = 3\ \text{g/cm}^3 \times 3750\ \text{cm}^3$ **(1)** $= 11\,250\ \text{g}$ **(1)** $= 11.25\ \text{kg}$ **(1)**

28   Investigating density

1   (a) mass **(1)**

    (b) electronic balance **(1)**

2   (a) For regularly shaped solids: any one of the following methods:

      1.   Volume can be directly measured using Vernier callipers / ruler to measure the length, width and height of the object **(1)**. The measurements are then multiplied together to find the volume, e.g. 3 cm $\times$ 3 cm $\times$ 3 cm to give the volume of a 3-cm cube **(1)**.

      2.   If the object is a regular shape, e.g. cube, cylinder, sphere, prism, the appropriate mathematical expression can be used **(1)**, e.g. use $4 / 3\ \pi\ r^3$ to find the volume of a sphere **(1)**.

      3.   If the density and mass are already known the volume can be calculated by using the equation **(1)**, i.e. volume $=$ mass / density **(1)**.

    (b) For an irregular solid: any one of the following methods:

      1.   Pour water into a measuring cylinder to a specific level and record the level **(1)**. Add the object to the water and record the new water level. The difference between the new water level and the original level will be the object's volume **(1)**.

      2.   Use a Eureka can by filling it with water until the water runs out from the spout **(1)**. When no more water runs out, carefully place the irregular solid into the can and measure the volume of water displaced through the spout by collecting the water in a measuring cylinder **(1)**.

3   (a) Place a measuring cylinder on a balance and then zero the scales with no liquid in the measuring cylinder **(1)**. Add the liquid and measure the level **(1)**. Record the mass of the liquid (in g) from the balance and the volume (in cm^3) by reading from the level in the measuring cylinder **(1)**.

    (b) Take the value at the bottom of the meniscus **(1)** making sure that the reading is made at 'eye level', to avoid a 'parallax error' **(1)**.

    (c) Density $=$ mass / volume so 121 g $\div$ 205 cm^3 **(1)** $= 0.59$ **(1)** g/cm^3

29   Changes of state

1   In a liquid there are some intermolecular forces between particles as they move round each other **(1)**. In a gas there are almost no intermolecular forces as the particles are far apart **(1)**.

2   Particles have different amounts in the kinetic energy store **(1)** and experience different intermolecular forces **(1)**.

3   B **(1)**

4   At boiling point the liquid changes state **(1)** so the energy applied after boiling point is reached goes into breaking bonds **(1)** between the liquid particles. The particles gain more energy and become a gas **(1)**.

5   The kinetic energy **(1)** of the particles decreases **(1)** as the ice continues to lose energy to the surroundings; this is measured as a fall in temperature **(1)**.

30   Internal energy

1   B **(1)**

2   At boiling point / latent heat of vaporisation there will be a change / increase to the potential energy of the particles **(1)** but the kinetic energy of the particles will not change **(1)**.

3   (a) When temperature rises due to heating, internal energy increases **(1)** because the kinetic energy of the particles increases **(1)**.

(b) When temperature does not rise, due to heating, internal energy increases **(1)** because the potential energy of the particles increases **(1)**.

4   When the water vapour condenses into liquid water there will be no change in the kinetic energy **(1)** of the water particles so the temperature does not change **(1)** but there will be a change in the potential energy **(1)** of the water particles as they move from a gas state to a liquid state.

31 Specific latent heat

1   Specific latent heat is the energy that must be transferred to change 1 kg of a material from one state of matter to another. **(1)**

2   C **(1)**

3   $E = m\,L = 25\,\text{kg} \times 336\,000\,\text{J/kg}$ **(1)** $= 8400\,000$ **(1)** J

4   (a) Melting – B **(1)**

   (b) Boiling – D **(1)**

   (c) Specific latent heat of fusion – B **(1)**

   (d) Specific latent heat of vaporisation – D **(1)**

   (e) The energy being transferred to the material is breaking bonds **(1)**; as a result, the material undergoes a phase change **(1)**.

5   $E = m\,L = 36\,\text{kg} \times 2260\,\text{kJ/kg}$ **(1)** $= 81\,360$ **(1)** kJ

32 Particle motion in gases

1   Temperature is a measurement of the average kinetic energy of the particles in a material. **(1)**

2   273K → 0 °C **(1)**

   255K → –18 °C **(1)**

   373K → 100 °C **(1)**

3   At a constant volume, the pressure and temperature of a gas are directly proportional. **(1)**

4   (a) As the temperature increases the particles will move faster **(1)** because they gain more energy **(1)**.

   (b) As the particles are moving faster they will collide with the container walls more often **(1)**, therefore increasing the pressure **(1)**.

   (c) It increases **(1)**

33 Pressure in gases

1   When particles of a gas collide **(1)** with a surface they exert a force at right angles to the surface **(1)** resulting in pressure **(1)**.

2   C **(1)**

3   $P_1 V_1 = P_2 V_2$ so $100\,\text{kPa} \times 230\,\text{cm}^3 = 280\,\text{kPa} \times V_2$. **(1)** $V_2 = 100\,\text{kPa} \times 230\,\text{cm}^3 / 280\,\text{kPa}$ **(1)** $= 82.1$ **(1)** cm^3

4   $P_1 = P_2 V_2 / V_1$ **(1)** $= 640\,\text{litres} \times 100\,\text{kPa} / 8\,\text{litres}$ **(1)** $= 8000\,\text{kPa}$ (or 8 MPa or 8×10^6 Pa) **(1)**

34 Extended response – Particle model

The answer should include some of the following points: **(6)**

- Solid, liquid and gas states of matter have increasing kinetic energy of particles.
- Thermal energy input or output will result in changes to the thermal energy store of the system and will result in changes of state or a change in temperature.
- Changes in states of matter are reversible because the material recovers its original properties if the change is reversed.
- Thermal energy input does not always result in a temperature rise if the energy is used to make or break bonds between particles / result in a change of state.
- Latent heat is the amount of heat / thermal energy required by a substance to undergo a change of state.
- The thermal energy required to change from solid / ice to water (accept converse) is called the latent heat of fusion and is calculated using $Q_f = ml$.
- The thermal energy required to change from liquid to gas / water to steam (accept converse) is called the latent heat of vaporisation and is calculated using $Q_v = ml$.

35 The structure of the atom

1   (a) Protons – labelled in the nucleus (+ charge) **(1)**

   (b) Neutrons – labelled in nucleus (0 charge) **(1)**

   (c) Electrons – labelled as orbiting (- charge) **(1)**

2   (a) The number of positively charged protons **(1)** in the nucleus is equal to the number of negatively charged electrons **(1)** orbiting the nucleus.

   (b) The atom will become a positively charged ion / charge of +1. **(1)**

3   Size of an atom: 10^{-10} m **(1)**

   Size of a nucleus: 10^{-15} m **(1)**

4   When an electron absorbs electromagnetic radiation **(1)** it will move to a higher energy level **(1)**. When the electron moves back from a higher energy level to a lower energy level **(1)** it will emit electromagnetic radiation **(1)**.

36 Atoms, isotopes and ions

1   (a) The name given to particles in the nucleus. **(1)**

   (b) The number of protons in the nucleus. **(1)**

   (c) The total number of protons and neutrons in the nucleus. **(1)**

2   C **(1)**

3   Isotopes will be neutral because the number of positively charged protons **(1)** still equals the number of negatively charged electrons **(1)**.

4   They both have 8 protons / they have the same proton number / atomic number **(1)**, and they both have 8 electrons **(1)** orbiting the nucleus. They have different numbers of neutrons: the first has 8 neutrons whereas the second has 10 **(1)**. (both needed for second mark)

5   Any two of the following explanations:

   1) An atom can lose one or more electrons by friction **(1)** where contact forces rub electrons away **(1)** from the atom **(1)**.

   2) An atom can lose one or more electrons by ionising radiation **(1)**, where electrons are removed from the atom by an alpha or beta particle colliding **(1)** with an electron.

   3) An atom or molecule can lose one or more electrons by electrolysis **(1)** when it was previously bonded in an ionic compound and is separated in solution in an electrolytic cell **(1)**.

37 Models of the atom

1   The plum pudding model showed the atom as a 'solid', positively charged **(1)** particle containing a distribution of negatively charged electrons **(1)** whereas the Rutherford model showed the atom as having a tiny, dense, positively charged nucleus **(1)** surrounded by orbiting negatively charged electrons **(1)**.

2   Rutherford fired positively charged alpha particles at atoms of gold foil and most went through, showing that most of the atom was space / a void **(1)**. Some were repelled or deflected **(1)**, showing that the nucleus was positively charged **(1)**.

3   (a) A **(1)**

   (b) The Bohr model showed that electrons orbited the atom at specific energy levels **(1)** and those electrons had to acquire precise amount of energy to move up to higher levels **(1)**. The 'excited' electrons also emitted discrete amounts of energy to move to a lower level **(1)**. The model was an improvement because it could explain emission and absorption spectra **(1)**.

38 Radioactive decay

1   (a) Activity is the rate **(1)** at which the unstable / radioactive nuclei decay per second **(1)**.

   (b) The unit of activity is the becquerel (Bq). **(1)**

   (c) Count rate is the number of counts of radioactive decay **(1)** per unit of time / second / minute. **(1)**

2   D **(1)**

3   In 1 second, 450 radioactive nuclei will decay **(1)** so $450 \times (2 \times 60)$ **(1)** $= 54\,000$ **(1)** nuclei in 2 minutes.

4   (a) Alpha radiation / particle: **(1)** the alpha particle consists of 4 nucleons / 2 protons and 2 neutrons / a 'helium' nucleus **(1)**

   (b) Beta radiation / particle: **(1)** a neutron changes to a proton increasing the positive charge by 1 **(1)**

39 Nuclear radiation

1   B **(1)**

2   alpha – very low, stopped by 10 cm of air

   beta minus – low, stopped by thin aluminium

   gamma – very high, stopped by very thick lead

   all correct, **(2)** 2 correct **(1)**

3   (a) No change in relative atomic mass. **(1)**

   (b) High-energy electron emitted from the nucleus. **(1)**

   (c) Moderately ionising. **(1)**

4   Compared with other types of ionising radiation, the chance of collision with air particles at close range is high **(1)** because the alpha particles have a large positive charge / are massive compared with other types of radiation **(1)**. Once an alpha particle has collided with another particle it loses its energy **(1)**.

5   Alpha and beta particles and gamma waves lose their energy when they collide **(1)** with shielding atoms, causing them (instead of body atoms) to become ionised **(1)**. The denser the shielding material the greater the chance of collision **(1)** and subsequent reduction in energy of the harmful radiation **(1)**.

40   Uses of nuclear radiation

1   (a) Beta **(1)** radiation is used because alpha radiation / particles **(1)** would not pass through and gamma radiation / waves / rays **(1)** would pass too easily.

   (b) (i)   The paper has become too thick. **(1)**

       (ii) The pressure on the rollers would be increased to make the paper thinner. **(1)**

2   They have high frequency / they carry large amounts of energy. **(1)**

3   (a) Alpha particles cannot pass through to the outside of the smoke alarm **(1)** and they are contained in a metal box / stopped by about 10 cm of air / are situated away from normal traffic of people **(1)**.

   (b) The smoke particles absorb the alpha particles **(1)** so the current falls / is broken and this triggers the bell to ring **(1)**.

4   Plastic instruments cannot always be heated to sterilise them **(1)** so gamma-rays can be used to kill bacteria/microbes **(1)**.

41   Nuclear equations

1   (a) α **(1)**

   (b) $\beta-$ **(1)**

2   B **(1)**

3   (a) (for nitrogen) 7 **(1)**

   (b) (for magnesium) 12 **(1)**

4   (a) beta-plus (positron) **(1)**

   (b) alpha particle **(1)**

   (c) neutron **(1)**

5   (a) add 208 to Po **(1)**; type of decay = alpha **(1)**

   (b) add 86 to Rn **(1)**; type of decay = alpha **(1)**

   (c) add 42 to Ca **(1)**; type of decay = beta-minus **(1)**

   (d) add 9 to Be **(1)**; type of decay = neutron **(1)**

42   Half-life

1   Half-life is the time taken for half the nuclei in a radioactive isotope to decay. **(1)**

2   (a) 8 million atoms **(1)**

   (b) 9.3 minutes = 3 half-lives **(1)** so, after 1 half-life, 8 million nuclei left, after 2 half-lives, 4 million nuclei left, and, after 3 half-lives, 2 million nuclei left **(1)**

3   (a) The activity is 400 Bq at 1.5 minutes **(1)** (between 1.3 and 1.7 is allowed). Half this activity is 200 Bq, at 6.5 **(1)** minutes (between 6.3 and 6.7 is allowed), so the half-life is 6.5 min to 1.5 min = 5 min **(1)**. (Answers between 4.7 and 5.3 min are allowed.) *(If you used other points on your graph and got an answer of around 5 min you would get full marks. For this question your working can just be pairs of lines drawn on the graph.)*

   (b) Net decline tells you what ratio **(1)** of the radioactive material has decayed after each half-life **(1)**.

   (c) 7/8th of the radioactive material has decayed after three half-lives. **(1)**

43   Contamination and irradiation

1   Before 1920, the effects of radium were not known / recognised **(1)** so it was thought that it was safe to use **(1)**. It was banned from use once the dangers were known **(1)**.

2   (a) External contamination: radioactive particles come into contact with skin, hair or clothing. **(1)**

   (b) Internal contamination: a radioactive source is eaten, drunk or inhaled. **(1)**

   (c) Irradiation: a person becomes exposed to an external source of ionising radiation. **(1)**

3   (a) Any suitable example, e.g. contaminated soil may get on to hands. **(1)**

   (b) Any suitable example, e.g. contaminated dust or radon gas may be inhaled. **(1)**

4   Internal contamination means that the alpha particles come into contact with the body through inhalation or ingestion **(1)**, where they are likely to cause internal tissue damage **(1)**. Alpha particles that are irradiated are less likely to cause damage because they have to travel through air **(1)** and are therefore less likely to ionise body cells **(1)** (at distances of over 10 cm).

44   Hazards of radiation

1   A **(1)**

2   1. Limiting the time of exposure / keep the time that a person needs to be in contact with the ionising source as low as possible. **(1)**

   2. Wear protective clothing / wearing a lead apron will absorb much of the ionising radiation. **(1)**

   3. Increasing the distance from the radioactive source / the further a person is from the ionising radiation, the less damage it will do. **(1)**

3   A source of alpha particles with high activity inside the body will ionise body cells **(1)** because they are highly ionising / massive / undergo many collisions **(1)** before transferring all of their ionising energy. Gamma-rays can pass out of the body fairly easily **(1)** without causing much damage to cells **(1)**.

4   Radioactive tongs allow the source to be kept as far as possible away from a person's hand **(1)** and allows it to be pointed away from people at all times **(1)**.

5   Those who use X-rays on a regular basis, such as medical workers, would have a high exposure / dose of radiation, which would cause damage if the dose was too high **(1)**. They leave the room so that they are not exposed to high levels of cumulative radiation / high dose **(1)**. The number of X-rays that a patient has is carefully monitored to minimise risk of high dose / exposure to high radiation **(1)**.

45   Background radiation

1   Radon is a radioactive element **(1)** that is produced when uranium in rocks decays **(1)**.

2   Levels can vary because of the different rocks **(1)** that occur naturally in the environment. They can also vary due to the use of different rocks such as granite **(1)** in buildings.

3   Natural: air **(1)** and **two** of the following – cosmic rays, rocks in the ground, food **(1)**

   Manufactured: nuclear power, medical treatment, nuclear weapons **(1)** (all three needed for mark)

4   (a) (add the three values and divide by 3) SE 0.27 Bq, SW 0.30 Bq **(1)** (both needed for mark)

   (b) The South West (has the highest average level of background radiation) **(1)**

46   Medical uses

1   B **(1)**

2   Any **three** from the following: maximising the distance from the source of radiation; **(1)** using special tools such as tongs and gloves when handling the radiation **(1)**; minimising the time of exposure to the radiation **(1)**; shielding bodies from exposure to the radiation with thick concrete barriers **(1)**; shielding bodies from exposure to the radiation with thick lead plates / aprons **(1)**

3   A medical tracer is a radioactive solution that contains a gamma-emitting radioisotope **(1)**. It is injected into the patient and is then absorbed by the organ being examined **(1)**. A special camera detects the gamma radiation emitted by the solution **(1)**. The detected waves are used to build up an image of where the radioisotope is located in the organ **(1)**.

4   The radioactive isotope must have a half-life long enough to give a useful image **(1)**, but short enough so that its nuclei have mostly decayed after the image has been taken **(1)**.

47   Nuclear fission

1   Left nucleus: uranium-235 **(1)**. Two new nuclei: daughter nuclei **(1)**. Small single particles – neutrons **(1)**.

2   (a) Nuclear fission is the splitting **(1)** of a large and unstable nucleus (e.g. uranium or plutonium). **(1)**

   (b) Usually, for fission to occur the unstable nucleus **(1)** must first absorb a neutron **(1)**.

3   (a) A chain reaction occurs as neutrons are released from a fission reaction and are absorbed by more uranium nuclei that also undergo fission **(1)**. As more than one neutron is released per fission reaction **(1)** this leads to more and more fission reactions occurring which could lead to an explosion **(1)**.

(b) Control rods are used to absorb excess neutrons to keep the chain reaction at a steady rate. **(1)**

(c)

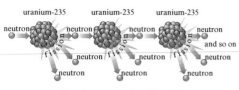

uranium-235   uranium-235   uranium-235

neutron   neutron   neutron   neutron

and so on

neutron   neutron   neutron

neutron   neutron   neutron

Uranium atom absorbing / about to absorb a neutron **(1)**. Three neutrons and energy released **(1)**. At least two more uranium atoms shown **(1)** each releasing three neutrons **(1)**.

(d) Controlled chain reactions are used in nuclear power stations. **(1)**

48   Nuclear fusion

1   B **(1)**

2   Fusion needs high pressures and temperatures **(1)**. At present, this requires more energy than is released by fusion **(1)**.

3   When fusion occurs the mass of the products is slightly less than the mass of the reactants **(1)** so the difference is released as energy in the form of light and heat **(1)**.

4   (a) Nuclei need to get very close to each other before fusion can happen **(1)**. Electrostatic repulsion means that the positive charges of the nuclei repel each other **(1)**.

(b) To give the nuclei enough energy **(1)** to overcome this, very high temperatures **(1)** and pressures **(1)** are needed.

49   Extended response – Radioactivity

The answer should include some of the following points: **(6)**

- All three types of radiation can pass into / penetrate different materials.
- Alpha particles have high relative mass and so transfer a lot of energy when they collide, so they are good at ionising.
- Alpha particles produce a lot of ions in a short distance, losing energy each time.
- Alpha particles have a short penetration distance so are absorbed by low density / thin materials such as a few centimetres of air and a sheet of paper.
- Beta particles have a low relative mass and can pass into / through more materials than alpha particles.
- Beta particles are less ionising than alpha particles and can be absorbed by 3-mm-thick aluminium.
- Gamma waves are high-frequency EM waves and can travel a few kilometres in air.
- Gamma waves are weakly ionising and need thick lead or several metres of concrete to absorb them. **(6)**

50   Scalars and vectors

1   (a) Scalars: speed, energy, temperature, mass, distance **(1)**

Vectors: acceleration, displacement **(1)** force (or weight), velocity, momentum **(1)**

(b) Any correct choice and explanation, e.g. mass **(1)** is a scalar because it has a size / magnitude **(1)** but no direction **(1)**.

2   (a) (i) Velocity is used because both a size and a direction are given. **(1)**

(ii) The students are jogging in opposite directions so the negative sign for one student indicates this. **(1)**

(b) The length of the arrow is proportional to the magnitude of the vector (in this case velocity) **(1)**. An increase in velocity to 3 m/s would mean an arrow 1½ times longer than that for 2 m/s **(1)**.

3   (a) D **(1)**

(b) Weight has size / magnitude and direction but the other quantities just have a magnitude. **(1)**

51   Interacting forces

1   gravitational **(1)**, magnetic **(1)**, electrostatic **(1)**

2   A **(1)**

3   Weight and normal contact force are vectors because they have a direction **(1)**. Weight is measured downwards **(1)** whereas normal contact force is measured upwards/opposite to weight **(1)**.

4   (a) pull (by the student on the bag) and friction/drag of the bag against the floor **(1)**.

(b) weight and normal contact/reaction force **(1)**

5   As the skydiver leaves the plane, weight acting downwards is greater than air resistance acting upwards so he accelerates **(1)**. As speed increases, air resistance increases to become equal to weight, so there is no net force / terminal velocity reached **(1)**. When the skydiver opens the parachute, air resistance upwards is greater than weight downwards so he decelerates **(1)**. The skydiver decelerates until air resistance upwards equals weight downwards – there is no net force (so a new terminal velocity is reached) **(1)**. (or similar wording)

52   Gravity, weight and mass

1   (a) The mass of the LRV on the Moon is 210 kg **(1)** because the mass of an object does not change if nothing is added or removed. **(1)**

(b)

Arrow points vertically downwards / top of arrow estimated around the seat area of the LVR **(1)**. (both points needed for mark) The centre of mass is where the weight of a body can be assumed to act downwards through a single point **(1)**.

2   $W = m\,g$ **(1)** so $(1 + 2 + 1.5)$ kg $\times$ 10 N/kg = 4.5 kg $\times$ 10 N/kg **(1)** = 45 N

3   Calculating correct masses for all three items. **(1)**

Selecting correct items **(1)** (clothes 10.5 kg + camera bag 5.5 kg + jacket 3.5 kg) = 19.5 kg **(1)**

53   Resultant forces

1   (a) A: 9.5 N; **(1)** B: 2 N; **(1)** C: 4.5 N; **(1)** D: 12.75 N **(1)**

(b) A: up; **(1)** B: up; **(1)** C: to the right; **(1)** D: to the left **(1)**

2   B **(1)**

3   (a) diagonal arrow – 2.5 cm **(1)**

(b) 50 N **(1)**

4   Scale correct, e.g. vertical 2 cm (6 N) and horizontal 5 cm (15 N) **(1)**

Hypotenuse = 5.4 **(1)** cm; represents 16.2 **(1)** N.

54   Free-body force diagrams

1   B **(1)**

2   (a) Upward **(1)** and downward **(1)** – same length.

(b) Force upwards 20 N **(1)**

Weight downwards 20 N **(1)**

3   (one only for all arrows) **(1)**

Longest arrow to left indicating movement to left. **(1)**

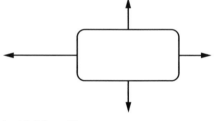

4   (a) 3.6 cm **(1)**

(b) 7.2 N **(1)**

55   Work and energy

1   A **(1)**

2   (a) Work done against friction will lead to a rise in temperature of the object **(1)** which is dissipated to the environment **(1)**.

(b) The greater the amount of friction, the more work that has to be done **(1)** to move the body through the same distance **(1)**.

3   $s = W / F$ so $s = 4800 / 80$ **(1)** = 60 **(1)** m

4   (a) $h = E / (m\,g)$ so $h = 320 / (8 \times 10)$ **(1)** (remember to convert the grams to kilograms) = 4 **(1)** m

(b) Use $W = F\,s$ so $F = W / s$ **(1)** and $F = 320 / 4$ **(1)** = 80 **(1)** N

56   Forces and elasticity

1   C **(1)**

2   (a) Tension: washing line (or any valid example) **(1)**

(b) Compression: G-clamp, pliers (or any valid example) **(1)**

(c) Elastic distortion: fishing rod (with a fish on the line) (or any valid example) **(1)**

(d) Inelastic distortion: dented can or deformed spring (or any valid example) **(1)**

3   After testing, Beam 1 would return to the same size and shape as before the test **(1)** and would be intact **(1)**. Beam 2 would distort and change shape **(1)** but would (probably) still be intact **(1)**.

4   Car manufactures install crumple zones / seat belts / air bags **(1)** in cars. These are parts of the car body that are designed to distort / change shape **(1)** in the event of a crash. They extend the time taken for a body to come to rest, reducing the force on the body **(1)**.

57 Force and extension

1   Elastic deformation means that the object will change shape in direct proportion to the force(s) exerted, up to the limit of proportionality **(1)**, and the change in shape is not permanent **(1)**. Inelastic deformation means that the object will change shape beyond the limit of proportionality / the limit of proportionality is exceeded **(1)** and the change in shape will be permanent **(1)**.

2   Extension = 0.07 m - 0.03 m = 0.04 m

Force = (spring constant / k) × extension = 80 N × 0.04 m **(1)**

Force = 3.2 **(1)** N **(1)**

3   D **(1)**

4   (a) $F = k e$ so $k = F / e$ **(1)** = 30 N / 0.15 m **(1)** = 200 **(1)** N/m

(b) $E = \frac{1}{2} k e^2 = \frac{1}{2} \times 200$ N/m × (0.15 m)^2 **(1)** = 2.25 J **(1)**

58 Forces and springs

1   (a) Hang a spring from a clamp attached to a retort stand and measure the length before any masses or weights are added using a half-metre ruler, marked in mm **(1)**. Carefully add the first mass or weight and measure the total length of the extended spring **(1)**. Unload the mass or weight and re-measure the spring to make sure that the original length has not changed **(1)**. Add at least five masses or weights and repeat the measurements each time **(1)**.

(b) The elastic potential energy can all be recovered **(1)** and is not transferred to cause a permanent change of shape in the spring **(1)**.

(c) Masses must be converted to force (N) by using $W = m \times g / F = m \times g$ **(1)**. The extension of the spring must be calculated for each force by taking away the original length of the spring from each reading **(1)**. Extension measurements should be converted to metres. **(1)**

(d) (i) The area under the graph equals the work done/the energy stored in the spring as elastic potential energy. **(1)**

(ii) The gradient of the linear part of the force–extension graph gives the spring constant k. **(1)**

(e) Limit of proportionality. **(1)**

(f) energy stored = $\frac{1}{2} \times k \times e^2$ **(1)**

2   The length of a spring is measured with no force applied to the spring whereas the extension of a spring is the length of the spring measured under load/force less the original length. **(1)**

59 Moments

1   B **(1)**

2   When an object is balanced the clockwise moment **(1)** is equal to the anticlockwise moment **(1)**.

3   $M = F d = 25$ N × 0.28 m **(1)** = 7 **(1)** N m **(1)**

4   (a) No **(1)**

Moment for Ben = 300 N × 0.8 m = 240 N m **(1)**

Moment for Amberley = 250 N × 1.2 m = 300 N m **(1)**

(b) Moment for Ben must change to equal 300 N m so distance must change to 300 N m / 300 N = 1 m **(1)**

(c) Original moment for Ben = 240 N m, so new moment for Amberley must equal this **(1)**

240 N m / 250 N = 0.96 m **(1)** (so Amberley must move to 0.96 m from the pivot)

60 Levers and gears

1   B **(1)**

2   (a) (i)   input forces as shown **(1)** (both needed)

(ii) output forces as shown **(1)** (both needed)

(iii) pivots as shown **(1)** (both needed)

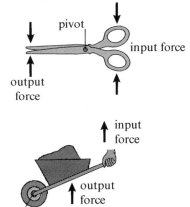

(b) The input force is the force provided by the user of the lever **(1)**. The output force is the force that results from the input force **(1)**.

3   For a high gear, the driver gear has a greater diameter than the driven gear **(1)** and the output is low for a given input force. **(1)** For a low gear, the driver gear has a similar diameter to the driven gear **(1)** and the output is high for a given input force **(1)**.

4   The cyclist would change to a low gear when moving from a horizontal road to a hill because a high output is needed **(1)** for a given force to move the bicycle up the hill against gravity **(1)**.

61 Pressure and upthrust

1   D **(1)**

2   $P = F / A$ so $P = 25 / 0.0625$ **(1)** = 400 **(1)** Pa

3   The weight of the box is found by $F = P A$ so $F = 2000$ Pa × 0.25 m^2 **(1)** = 500 N **(1)**

4   (a) The more cargo a ship carries, the deeper it sits in the water **(1)**. When ships are not carrying cargo, they weigh less and so displace less water **(1)** and the hull / Plimsoll line is higher above the surface of the water **(1)**.

(b) Warm seawater produces slightly less upthrust than cold seawater **(1)**, so ships will float lower down in warm water **(1)** and the Plimsoll line will be higher up the hull to avoid the ship becoming overloaded **(1)**.

62 Pressure in a fluid

1   Water pressure increases with depth **(1)** because there is a greater weight of liquid with increasing depth **(1)**.

2   (a) $P = h \rho g = 1500$ m × 1025 kg/m^3 × 10 N/kg **(1)** = 15 375 000 **(1)** Pa **(1)**

(b) Pressure at a point increases with the height of the column of liquid above that point **(1)** and with the density of the liquid **(1)**.

3   (a) Atmospheric pressure describes a column of air, reaching from the Earth's surface to the top of the atmosphere **(1)**, and covering one square metre of the Earth **(1)** containing a mass of air equal to 10 000 kg **(1)**.

(b) $P = 100 000$ Pa and $A = 1 \ m^2$ so force (weight) = 100 000 Pa × 1 m^2 = 100 000 N **(1)**

$F = m g$, or $m = F / g$ **(1)**

so $m = 100 000 / 10 = 10 000$ kg **(1)**

63 Distance and displacement

1   C **(1)**

2   Distance does not involve a direction and so is a scalar quantity **(1)**. Displacement involves both the distance an object moves and the direction that is has moved in from its starting point so it is a vector quantity **(1)**.

3   (a) The circumference of the Ferris wheel is $2 \pi \times (15 / 2) = 47.1$ m. **(1)** The wheel completed three cycles so the total distance travelled was 3×47.1 **(1)** = 141.3 m **(1)**. (accept 141 m)

(b) The final displacement was 0 m **(1)** as the girls returned to the starting point at the end of the ride. **(1)**

64 Speed and velocity

1   Average speed = 10 000 m / 2400 s = 4.17 **(1)** m/s

2   (a) Any **three** of the following: age, terrain, fitness, distance travelled. **(1)**

(b) (i)   walking: 1.5 m/s **(1)**

(ii) running: 3 m/s **(1)**

(iii) cycling: 6 m/s **(1)**

3   The speed of the satellite can be constant because the distance being covered each second is constant **(1)**, but the velocity changes constantly because direction of motion is constantly changing **(1)**.

4   (a) $s = v t$, so distance = $4 \times (15 \times 60)$ **(1)** = 3600 **(1)** m.

(b) The term velocity is used because the boat has both a speed **(1)** and a direction. **(1)**

(c) At the finish of the journey by boat, the displacement of the team from the boathouse is 3600 m **(1)**. As they return to the boathouse by bus, the displacement decreases **(1)** until they arrive back at the boathouse where the final displacement is 0 m **(1)**.

65 Distance–time graphs

1. (a) (i) B **(1)**

   (ii) C **(1)**

   (b) Evidence of attempt to calculate the gradient of the slope to find speed / use of change in distance divided by the change in time / speed = distance / time, **(1)** so s = 20 / 40 **(1)** = 0.5 **(1)** m/s.

   (c) Walking (slowly) **(1)** as the average walking speed is 1.5 m/s **(1)**.

   (d) In part A, he travels 60 m in 40 s. Speed = distance / time **(1)** = 60 m / 40 s **(1)** = 1.5 **(1)** m/s. (You can use any part A of the graph to read off the distance and the time as the line is straight; you should always get the same speed.)

   (e) Displacement is the length and direction of a straight line between the runner's home and the park **(1)**, but the distance the runner ran may have included bends and corners on the path that the runner took **(1)**.

66 Velocity–time graphs

1. (a) $a = \Delta v / t = 4 / 5$ **(1)** $= 0.8$ **(1)** m/s^2 **(1)**

2. (a) A **(1)**

   (b) A right-angled triangle with a horizontal side and a vertical side that covers as much of the line as possible for precision. **(1)**

   (c) Change in velocity = 30 - 0 m/s. Time taken for change = 5 - 0 s.

   Acceleration = $\dfrac{\text{change in velocity}}{\text{time taken}}$

   or $a = \Delta v / t = 30$ m/s / 5 s **(1)** = 6 **(1)** m/s^2. (The triangle drawn may be different but the answer should be the same.)

   (d) Area under line or distance = ½ × 5 s × 30 m/s **(1)** = 75 **(1)** m

67 Equations of motion

1. Attempt to estimate area under the graph / count squares **(1)** = approximately 21 squares or ~ 21 **(1)** m. (allow 20–22)

2. (a) $a = \Delta v / t$ **(1)** = (25 m/s – 15 m/s) / 8 s **(1)** = 1.25 **(1)** m/s^2

   (b) $v^2 = u^2 + 2as = (25$ m/s$)^2 + 2(1.25$ m/s$^2 ×$ 300 m) **(1)** = 1375 m/s^2 **(1)**

   $v = \sqrt{1375}$ m/s = 37 m/s **(1)** (allow rounding error – answers between 37.00 m/s and 37.10 m/s)

   (c) $v^2 – u^2 = 2\,a\,s$, so $s = (v^2 – u^2) / (2\,a)$ **(1)** = ((5 m/s)^2 – 1375 m/s^2) / (2 × –2 m/s^2) = **(1)** = –1350 m/s^2 / –4 m/s^2

   Distance = 337.5 m **(1)**

68 Terminal velocity

1. C **(1)**

2. (a) The force of gravity pulls the skydiver downwards **(1)**. Air resistance is small so the resultant force is large **(1)**.

   (b) The skydiver stops accelerating because the air resistance becomes equal to her weight **(1)** so the resultant force is zero **(1)**.

   (c) Terminal velocity **(1)**

   (d) The larger surface area increases air resistance **(1)** so the resultant force acts upwards **(1)** slowing the skydiver down **(1)**.

3. Ben should measure and mark the column with equal distances (e.g. every 20 cm) **(1)**. The ball should then be dropped through the viscous liquid and the time recorded between each marker **(1)**. When the ball passes through two marked distances at the same speed, terminal velocity is reached **(1)**. Ben can then use $s = d / t$ to calculate the speed at which terminal velocity occurs **(1)**.

69 Newton's first law

1. Four arrows drawn: vertical: down = weight, up = upthrust (arrows the same length); horizontal left to right = driving force from the engines, right to left = water resistance or resistive force; the driving force arrow should be longer than the resistive force arrow. (1 mark for all the forces correctly named and 1 mark for all the corresponding relative lengths of the arrows)

2. (a) Resultant force = 30 N + (–5 N) + (–1 N) **(1)** = 24 **(1)** N

   (b) The resultant force is zero / 0 N **(1)** so the velocity is constant / stays the same **(1)**.

3. (a) Assume downwards is positive, so resultant downward force is positive = 1700 N – 1900 N **(1)** = –200 **(1)** N. (You should state which direction you are using as the positive direction.)

   (b) The velocity of the probe towards the Moon will decrease **(1)** because the force produces an upward acceleration / negative acceleration **(1)**.

   (c) After landing, the forces on the probe are balanced so there is zero resultant force / the probe does not move **(1)**. The probe will not move unless it is acted on by another force **(1)**. The tendency of a body / the probe to remain at rest is called inertia **(1)**.

70 Newton's second law

1. (a) The trolley will accelerate **(1)** in the direction of the pull / force **(1)**.

   (b) The acceleration is smaller / lower **(1)** because the mass is larger **(1)**.

2. (a) $F = m\,a$ = 3000 kg × –3 m/s^2 **(1)** = –9000 **(1)** N

   (b) Backwards / in the opposite direction to the motion of the minibus. **(1)**

3. (a) $a = F / m$ = 10 500 N / 640 kg **(1)** = 16.4 **(1)** m/s^2

   (b) The mass of the car decreases, **(1)** so the acceleration will increase **(1)**.

71 Force, mass and acceleration

1. Electronic equipment is much more accurate **(1)** than trying to obtain accurate values for distance and time to calculate velocity, then calculate acceleration **(1)**, when using a ruler and a stopwatch. (Reference should be made to distance, time and velocity.)

2. Acceleration is inversely proportional to mass. **(1)**

3. Acceleration is the change in speed ÷ time taken so two velocity values are needed **(1)**; the time difference between these readings **(1)** is used to obtain a value for the acceleration of the trolley.

4. (a) For a constant slope, as the mass increases, the acceleration will decrease **(1)** due to greater inertial mass **(1)**.

   (b) Newton's second law, $a = F ÷ m$ **(1)**

5. An accelerating mass of greater than a few hundred grams can be dangerous and may hurt somebody if it hits them at speed **(1)**. Any two of the following precautions: do not use masses greater than a few hundred grams **(1)**, wear eye protection **(1)**, use electrically tested electronic equipment **(1)**, avoid trailing electrical leads **(1)**.

72 Newton's third law

1. D **(1)**

2. As the rocket sits on the launch pad, its weight downwards is equal and opposite to **(1)** the reaction force upwards of the launch pad **(1)**, so the rocket does not fall through.

3. The weight of the penguin is pushing down on the ice and the reaction force of the ice is pushing back on the penguin **(1)**, so the penguin is supported by the ice and does not fall through **(1)**.

4. Newton's third law says that the force must be equal in size/magnitude **(1)** and opposite **(1)** in direction for equilibrium. The force exerted by the buttresses on the building **(1)** is equal and opposite to the force exerted on the buttresses by the building **(1)** resulting in no movement occurring.

73 Stopping distance

1. (a) Thinking distance + braking distance = overall stopping distance **(1)**

   (b) Speed increases by 3 times so thinking distance increases by 3, and therefore thinking distance = 3 × 6 m = 18 m. **(1)** Speed increases by 3 times so braking distance increases by 9 times and braking distance = 9 × 6 m = 54 m. **(1)** Overall stopping distance = 18 m + 54 m = 72 m **(1)**

   (c) Thinking distance will increase if: the car's speed increases, the driver is distracted, the driver is tired, or the driver has taken alcohol or drugs **(1)**. (All four points needed for mark.)

   Braking distance will increase if: the car's speed increases, the road is icy or wet, the brakes or tyres are worn, or the mass of the car is bigger **(1)**. (All four points needed for mark.)

2. $F\,d = ½\,m\,v^2$, so $F = (½\,m\,v^2) / d$, so $F = [½ × 1500$ kg × (8 m/s × 8 m/s)] / 75 m **(1)** = (750 kg × 64 m/s) / 75 m **(1)** = 640 **(1)** N

3. Driving faster will increase thinking distance **(1)** and braking distance **(1)**. If drivers do not increase their normal distance behind the vehicle in front accordingly, there is an increased risk of an accident / collision **(1)**.

74 Reaction time

1. B **(1)**

2. Human reaction time is the time taken between a stimulus occurring and a response **(1)**. It is related to how quickly the human brain can process information and react to it **(1)**.

3. (a) A person waits with his index finger and thumb opened to a gap of about 8 cm **(1)**. A metre ruler is held, by a partner,

so that it is vertical and exactly level with the person's finger and thumb / with the lowest numbers on the ruler by the person's thumb **(1)**. The ruler is dropped and then grasped by the other person as quickly as possible **(1)**.

(b) 0.2 s to 0.9 s **(1)**

(c) The distance measured on the ruler would be short for the person with a reaction time of 0.2 s **(1)** and longer for the person with a reaction time of 0.9 s **(1)**.

75 Momentum

1 The momentum of the car would change if it accelerates / speeds up or decelerates / slows down **(1)**. The momentum of the car would also change if it changed direction **(1)** because velocity is a vector **(1)**.

2 $\rho = m \, v$ so $\rho = 1200 \text{ kg} \times 30 \text{ m/s}$ **(1)**. momentum = 36 000 kg m/s **(1)** in the south / southerly direction **(1)**.

3 (a) Momentum of Dima and his car = 900 kg × 1.5 m/s **(1)** = 1350 **(1)** kg m/s

(b) (i) Momentum of Sam and car = 900 kg × 3 m/s = 2700 **(1)** kg m/s

(ii) The total momentum is conserved / total momentum of both cars is unchanged. **(1)**

(iii) Total momentum = 1350 + 2700 = 4050 kg m/s **(1)** so velocity after collision = total momentum / total mass = 4050 kg m/s / 1800 kg **(1)** = 2.25 m/s **(1)**

4 Momentum of skater 1 before collision = 50 kg × 7.2 m/s = 360 kg m/s; momentum of skater 2 before collision = 70 kg × 0 m/s = 0 kg m/s, so momentum of both skaters after collision = 360 + 0 = 360 kg m/s **(1)**, so combined velocity = 360 kg m/s / (70 + 50) kg **(1)** = 3 **(1)** m/s.

76 Momentum and force

1 Force is the rate of change of momentum. **(1)**

2 (a) $p = m \, v$ **(1)** = 1500 kg × 25 m/s **(1)** = 37 500 kg m/s **(1)**.

(b) $F = m \, \Delta v / \Delta t$ (to find Δv use answer from (a): Δv = 25 m/s) so F = 1500 kg × 25 m/s / 1.8 s **(1)** = 20 833 **(1)** N

3 (a) The forces exerted on the passenger are large when the mass **(1)** or the deceleration **(1)** of the vehicle are large.

(b) An air bag / crumple zone / seat belt **(1)** increases the time over which a passenger comes to rest so reducing the force exerted on them **(1)**.

4 $F = m \, \Delta v / \Delta t$ so F = 500 kg × 5 m/s / 20 s **(1)** = 125 **(1)** N

77 Extended response – Forces

The answer should include some of the following points: **(6)**

- Acceleration is the rate of change of velocity (speed in a given direction) so although the speed is constant, the direction is continually changing for an object in circular motion.
- For motion in a circle there must be a resultant force known as a centripetal force that acts towards the centre of the circle.
- The string provides the centripetal force which acts towards the centre of the circle.

- Extend the investigation with different lengths of string.
- Extend the investigation with different masses.
- Improve data collection with electronic sensors.
- Improve data collection with video analysis.
- Reference to the importance of control variables for valid data collection.

78 Waves

1 When energy travels through water, we can see that the water particles themselves do not travel. This is because an object on the surface will 'bob' up and down / vertically **(1)** as the wave passes horizontally **(1)**.

2 When energy travels through air, we can see that the sound is generated by a vibrating / oscillating surface **(1)** (e.g. a loudspeaker) that causes air particles to vibrate / oscillate in the same plane, creating pressure waves / areas of compression and rarefaction / sound waves **(1)**.

3 (a) B **(1)**

(b) 12 cm / 2 (as there are 2 waves shown) so wavelength = 6 cm = 0.06 m **(1)**

(c) Any correct wave with higher amplitude (height) **(1)** and shorter wavelength **(1)**.

(d) To find frequency use $v = f \lambda$, so f = 3 × 10^8 m/s / 0.12 m **(1)** = 2 × 10^9 Hz **(1)** (or 2 × 10^6 kHz / 2500 MHz / 2.5 GHz)

To find time period use *period* = 1 / *frequency* so *period* = 1 / 2 × 10^9 **(1)** (allow error carried forward from finding frequency) = 5 × 10^{-10} seconds / s **(1)**

79 Wave equation

1 (a) $v = f \lambda$ rearranged is $f = v / \lambda$, so f = 1500 m/s / 88 m **(1)** = 17 **(1)** Hz

(b) $\lambda = v / f$ so λ = 1500 m/s / 22 Hz **(1)** = 68.2 m **(1)**

2 Wave speed = 0.017 m × 20 000 Hz **(1)** = 340 **(1)** m/s

3 $\lambda = v / f$ = 0.05 m/s / 2 **(1)** = 0.025 **(1)** m **(1)**

4 First, calculate the velocity of the waves $s = d / t$ **(1)** = 300 000 000 m/s (or 3 × 10^8 m/s) **(1)**, then use your calculated velocity to find the frequency using $v = f \lambda$, so $f = v / \lambda$ or $f = (3 \times 10^8) / 5$ **(1)** = 6 × 10^7 **(1)** Hz.

80 Measuring wave velocity

1 Frequency (*f*) = 3 Hz, wavelength (λ) = 0.05 m, speed of waves = 3 Hz × 0.05 m **(1)** = 0.15 **(1)** m/s **(1)**

2 D **(1)**

3 T = 4 divisions × 0.005 ms = 0.02 s **(1)**

period = 1 / *frequency* so *period* = 1 / 0.02 s **(1)** = 50 Hz **(1)**

81 Waves in fluids

1 (a) Count the number of waves that pass a point each second and do this for one minute **(1)**; divide the total by 60 to get a more accurate value for the frequency of the water waves **(1)**.

(b) Use a stroboscope to 'freeze' the waves **(1)** and find their wavelength by using a ruler in the tank/on a projection **(1)**.

(c) wave speed = frequency × wavelength or $v = f \times \lambda$ **(1)**

(d) the depth of the water **(1)**

---

2 A ripple tank can be used to determine a value for the wavelength, frequency and wave speed of water waves, **(1)** as long as small wavelengths **(1)** and small frequencies are used **(1)**.

3 water: hazard – spills may cause slippages; safety measure – report and wipe up immediately **(1)**; electricity: hazard – may cause shock or trailing cables; safety measure – do not touch plugs/wires/switches with wet hands or keep cables tidy **(1)**; strobe lamp: hazard – flashing lights may cause dizziness or fits; safety measure – check that those present are not affected by flashing lights **(1)**

82 Waves and boundaries

1 (a) C **(1)**

(b) Stone is more dense **(1)** and, the greater the difference in density between materials, the more sound energy will be reflected **(1)**.

2 Light waves from an object (on the surface) are reflected at a plane surface **(1)**. The angle of incidence equals the angle of reflection **(1)** when measured against a normal drawn perpendicular to the plane surface **(1)**. (Both required for mark.) This is repeated at the bottom of the periscope so that the light reaches the user **(1)** (or similar wording).

3 (a)

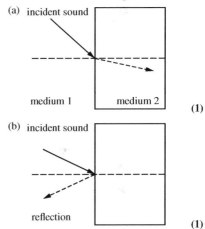

medium 1   medium 2   **(1)**

(b)

reflection   **(1)**

4 Sound is mostly transmitted through the boundary from one material to another when their densities are similar **(1)** but some refraction will occur **(1)**. Sound is mostly reflected at the boundary between a low-density material and a relatively high-density material **(1)**.

83 Investigating refraction

1 (a) Place a refraction block on white paper and connect a ray box to an electricity supply **(1)**; switch on the ray box and set it at an angle to the surface of the block **(1)**; use a sharp pencil to draw around the refraction block and make dots down the centre of the rays either side of the block **(1)**; use a sharp pencil and ruler to join the 'external' rays and then draw a line across the outline of the block to join the lines **(1)**; use a protractor to draw a normal where the light ray met the block and measure the angle of incidence and angle of refraction **(1)**.

(b) When a light ray travels from air into a glass block, its direction changes **(1)** and the angle of refraction will be less than the angle of incidence **(1)**.

(c) (i) The ray of light would not change direction. **(1)**

(ii) The light would slow down **(1)** (travelling from a less dense medium to a more dense medium) and the wave fronts would be closer together/ the wavelength of light would be shorter **(1)**.

2 any four from: use of electricity: if mains electricity is used there is a risk of shock – use tested apparatus/do not try to plug in/ unplug in the dark **(1)**; experiments are generally done in low-level light so there is a risk of tripping – clear floor area and working space (no trailing wires) and avoid moving around too much **(1)**; if glass blocks are used there is a risk of cuts – handle with care, use Perspex/non-breakable blocks when possible **(1)**; ray boxes get hot so risk of burns – do not touch ray boxes during operation **(1)**.

3 The waves travel more slowly and wavelength becomes shorter in shallow water **(1)**; the waves change direction/bend towards the 'normal' as they move into shallower water **(1)**.

84 Sound waves and the ear

1 B **(1)**

2 The particles are much further apart in air than in water **(1)** so it takes longer for vibrations to be passed from one particle to another **(1)**.

3 (a) A sound wave causes vibration of the eardrum tissue **(1)**. The vibration is transferred to the three 'solid' bones in the middle ear **(1)**. The vibration than passes through the bones (hammer, stirrup, anvil / causes vibrations that move through fluid in the cochlea / inner ear) **(1)**.

(b) The vibrations in the cochlea / inner ear cause small hair cells to move **(1)** and transform them into electrical impulses which are sent to the brain **(1)**.

4 In the human ear, the eardrum will not vibrate **(1)** if the wave frequency is less than 20 Hz or more than 20 kHz **(1)**. If there is no vibration of the eardrum, no sound is heard. **(1)**

85 Uses of waves

1 C **(1)**

2 As ultrasound waves pass into the body, some waves are reflected each time they meet a layer of tissue **(1)** with a different density **(1)**. The scanner detects the echoes **(1)** and the computer uses the information to make a picture **(1)**.

3 (a) The Earth's mantle becomes denser with increasing depth **(1)**; the wave speed depends on the density **(1)**, which depends on the increasing pressure **(1)**.

(b) $\lambda = v / f = 7$ m/s / 0.05 Hz **(1)** = 140 **(1)** m

4 Distance travelled by ultrasound = speed × time **(1)** = 8400 m/s × 0.5 × 10^{-9} s **(1)** = 4.2 × 10^{-6} m **(1)**

Distance of layer below the surface = ½ × 4.2 × 10^{-6} = 2.1 × 10^{-6} **(1)** m

86 Electromagnetic spectrum

1 (a) All waves of the electromagnetic spectrum are transverse waves **(1)** and they all travel at 3×10^8 m/s / the same speed in a vacuum **(1)**.

(b) Electromagnetic waves all transfer energy from the source of the waves to an absorber. **(1)**

2 (a) A: X-rays **(1)**

B: visible light **(1)**

C: microwaves **(1)**

(b) Frequency increases from radio to gamma waves **(1)**. The energy of the waves increases with frequency **(1)**, so gamma-rays have the most energy and radio waves the least **(1)**.

3 $v = f\lambda$, so $f = v / \lambda$ **(1)** = 3×10^8 m/s / 240 m **(1)** = 1.25×10^6 **(1)** Hz

87 Properties of electromagnetic waves

1 C **(1)**

2 (a) Reflection – waves bounce off a surface; **(1)** refraction – waves change speed and direction when passing from one material to another **(1)**; transmitted – electromagnetic waves are transmitted when they pass through a material **(1)**; absorbed – different electromagnetic waves are absorbed by different materials **(1)**.

(b) Any **two** valid examples: e.g. reflection – light on a mirror **(1)**; refraction – light through water **(1)**; transmission – radio waves passing through the atmosphere from transmitter to receiver / X-rays absorbed by the atmosphere **(1)**.

3 (a) Microwaves are shorter in wavelength **(1)** and higher in frequency **(1)** than radio waves.

(b) Microwaves sent from the transmitter are able to pass through the ionosphere **(1)** and are received and re-emitted by the receiver to the ground **(1)**. Radio waves are sent from the transmitter but are refracted by the ionosphere **(1)** and then reflected back to the receiver on the ground **(1)**.

4 Space-based telescopes are outside the Earth's atmosphere **(1)** so they can detect the whole range of electromagnetic waves **(1)** that are emitted by stars and galaxies **(1)**.

88 Infrared radiation

1 (a) any four from: Fill Leslie's cube with hot water at a known temperature (e.g. wait until it falls to 80 °C before taking temperature measurements) **(1)**. Measure the temperature at a distance (e.g. 10 cm) **(1)** from one of the four sides of Leslie's cube for a period of time (e.g. 5 minutes) **(1)**. Take regular readings (e.g. every 30 seconds) **(1)**. Repeat method for the other three sides **(1)**.

(b) independent variable: sides of Leslie's cube; **(1)** dependent variable: temperature **(1)**

(c) Any four from: starting temperature of the water, distance of heat sensor/thermometer from the cube, length of time, same number of readings taken, same intervals of time for each temperature reading, use the same thermometer or temperature

sensor. All four correct for **2 marks**; three correct for **1 mark**; two or fewer correct for **0 marks**.

2 (a) and (b) Any one hazard and method of minimising it from: Hot water in eyes can cause damage **(1)** – always wear eye protection **(1)**. Boiling water can cause scalds **(1)** – place the kettle close to cube and fill the cube in its place **(1)**. The cube can cause burns **(1)** – do not touch the cube until the temperature reading is low. **(1)** Water and electricity can result in a shock **(1)** – when using an electrical temperature sensor, keep well away from water **(1)**. Trailing wires can be a trip hazard **(1)** – avoid trailing leads/tuck leads out of the way **(1)**.

3 (a) The bungs would minimise thermal energy transferred from the flasks through evaporation. **(1)**

(b) (i) Dull and black surfaces are the best emitters and best absorbers. **(1)**

(ii) Shiny and light surfaces are the worst emitters and worst absorbers. **(1)**

89 Dangers and uses

1 (a) Infrared waves: (any **two** from) night vision goggles / security sensor / TV remote control / cooking / thermal imaging camera **(1)**

(b) Ultraviolet waves: (any **two** from) disinfecting water / sterilising surgical / scientific instruments / entertainment lighting / security marking **(1)**

(c) Gamma waves: (any **two** from) sterilising food / treating cancer / detection of cracks (pipes / aircraft, etc.) **(1)**

(d) Communication / mobile phones / satellites **(1)**

2 (a) The oral X-ray delivers the least amount of radiation, 0.005 mSv, which is the same as 1 day's normal exposure to background radiation. The next higher dose is for a chest X-ray which delivers 0.1 mSv, equivalent to 10 days of background radiation, and the highest dose is for the lung cancer screening, which delivers 1.5 mSv, equivalent to 6 months' exposure to background radiation. (X-rays in the right order, **(1)** dose related to equivalent background radiation for all three, **(1)** data quoted from table for all three **(1)**))

(b) If a person has too many X-rays in a certain time, particularly higher-dose X-rays, they may be more at risk of damage to body cells **(1)** if the X-rays are not carefully controlled to allow the body to recover from the high-energy doses **(1)**.

3 Radio waves can be produced by oscillations in electrical circuits **(1)**. Radio waves can also induce oscillations **(1)** in electrical circuits by creating an alternating current **(1)** with the same frequency as the radio waves, when they are absorbed **(1)**.

90 Lenses

1 A converging lens bends the rays of light towards each other **(1)**. A diverging lens bends the rays of light away from each other **(1)**.

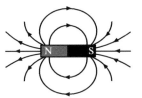

2 (a) If the lens is thicker, the focal point will be closer to the lens / shorter **(1)**. If the lens is thinner, the focal point will be further away from the lens / longer **(1)**.

(b) The focal length is the distance from the lens to the principal focus. **(1)**

3 (a) Thick lens: mid-ray continues unchanged; top and bottom rays angle down steeply **(1)**, focal point relatively close to the lens **(1)**.

Thin lens: mid-ray continues unchanged; top and bottom rays angle down less steeply than for lens 1 **(1)**, focal point further from lens **(1)**.

(b) Convex lens **(1)**

4 (a) Three construction rays **(1)**, divergent rays extrapolated **(1)**, focal point correctly identified **(1)**, focal length correctly labelled **(1)**.

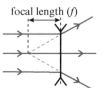
focal length (*f*)

(b) (i) The image would be virtual **(1)**

(ii) Concave lenses only form virtual images **(1)** because the rays diverge / are extrapolated by the brain of the observer. **(1)**

91 Real and virtual images

1 A real image is an image that can be projected on a screen **(1)** but a virtual image cannot be projected on a screen **(1)**. A virtual image is produced when light rays appear to come from a point beyond the object **(1)**.

2 (a) Points added at the distance 2F on both sides of the lens. **(1)**

(b) Point added at the same distance as F on the right hand side of the lens. **(1)**

(c) Vertical arrow drawn from where the two rays of light cross on the right-hand side up to the central line. **(1)**

3 Image height = magnification × object height, so image height = 4.5 × 3.3 **(1)** = 14.85 mm **(1)**

4 (a) The magnification of the image increases. **(1)**

(b) When the object is closer to the lens than the focal point. **(1)**

(c) At 2F **(1)**

92 Visible light

1 C **(1)**

2 Any plane surface such as a mirror, very smooth water, polished glass. **(1)**

3 An opaque green object appears green because it reflects green light **(1)** and all other colours are absorbed **(1)**.

4 (a) Violet light has a shorter wavelength than red light **(1)** so it carries more energy.

(b) Yellow light is a combination of red light and green light **(1)**. The green book cover absorbs red wavelengths of light but reflects green wavelengths of light, so it still looks green **(1)**. The blue title absorbs both red and green wavelengths of light and so looks black **(1)**.

5

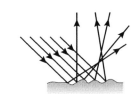

Incident rays of incidence all parallel, reflected rays reflected in different directions **(1)**. Uneven surface shown to represent rough surface **(1)**. Diffuse reflection still obeys the law of reflection because each ray of light that arrives at a surface is still reflected **(1)** according to the law of reflection (angle *i* = angle *r*) **(1)**. At the microscopic level, the surface is not smooth / the microscopic surfaces are at different angles to each other **(1)**.

93 Black body radiation

1 B **(1)**

2 Cup A will radiate more heat in 5 minutes **(1)** because the hotter the body the more infrared radiation it radiates in a given time **(1)**.

3 Maintenance of the average temperature of the Earth relies on a balance between absorption of energy from the Sun **(1)** and emission / radiation of energy into space **(1)**. Carbon dioxide absorbs thermal energy **(1)** and so reduces the amount of energy radiated into space. Scientists believe this leads to increased temperature / global warming of the Earth **(1)**.

4 Astronomers measure intensity of wavelengths of the light from a star to determine the peak / main wavelength **(1)**; they can then relate this to a scale of wavelength emitted at certain temperatures and so determine the temperature of the star **(1)**. (or similar wording)

94 Extended response – Waves

The answer should include some of the following points: **(6)**

- X-rays and gamma waves are both transverse waves.
- X-rays and gamma waves have a high frequency and therefore carry high amounts of energy.
- X-rays and gamma waves cause ionisation in atoms and exposure can be dangerous / cause cells to become cancerous.
- People who work regularly with X-rays and gamma waves should limit their exposure by using shields or leaving the room during use.
- X-rays are transmitted by normal body tissue but are absorbed by bones and other dense materials such as metals.
- X-rays and gamma waves mostly pass through body tissue but can be absorbed by some types of cells such as bone.
- X-rays and gamma waves can be used to investigate / treat medical problems.
- X-rays and gamma waves are used in industry to examine / 'see' into objects to examine for cracks / failures. **(6)**

95 Magnets and magnetic fields

1 Field lines **out** (arrows) of N **(1)**, field lines **in** (arrows) at S **(1)**, field line close at poles, **(1)** further apart at sides **(1)**.

---

2 A bar magnet and the Earth both have north and south poles **(1)**. A bar magnet and the Earth have similar magnetic field patterns **(1)**. The direction of both fields can be found using a plotting compass **(1)**.

3 An induced magnet is used for an electric doorbell because it can be magnetised when the current is switched on **(1)**, which attracts the soft iron armature to ring the bell **(1)**, and is de-magnetised when the current is switched off **(1)** (moving the armature away from the bell). An induced magnet is used as the 'switching' off and on of the magnet allows the arm to be moved and the bell rung **(1)**. Also, the default 'rest' state of the bell does not need any energy input **(1)**.

96 Current and magnetism

1 (a) At least two concentric circles on each diagram. **(2)**

(b) Clockwise arrows on cross diagram **(1)**; anticlockwise arrows on dot diagram **(1)**.

2 B **(1)**

3 The force acting on a current-carrying wire in a magnetic field depends on the length / *l* **(1)** of the wire, the current / *I* **(1)** in the wire and the magnetic flux density / *B* **(1)**.

4 (a) *F* = *B I l* so *F* = 0.005 × 1.4 × 0.56 **(1)** = 0.0039 **(1)** N

(b) *F* = 0.005 × 2.8 × 0.56 **(1)** = 0.0078 **(1)** N

(c) *F* = 0.005 × 1.4 × 0.23 **(1)** = 0.0016 **(1)** N

97 Current, magnetism and force

1 (a) When a current-carrying wire is placed in a magnetic field it would move **(1)**. This is because the current-carrying wire has its own magnetic field **(1)**, which interacts with the magnetic field / is repelled by the magnetic field / is attracted to the magnetic field **(1)**.

(b) D **(1)**

2 First finger – field **(1)**; second finger – current **(1)**; thumb – movement **(1)**.

3 The size of the force can be increased by increasing the strength of the magnetic field **(1)** or by increasing the current **(1)**.

4 *F* = *B I l* so *I* = *F* / *Bl* so *I* = 0.21 × 10^{-3} N / 0.0005 T × 0.30 m **(1)** so *I* = 1.4 **(1)** A **(1)**

98 The motor effect

1 Any one from: The wire can be coiled / turned into a solenoid **(1)**. The current can be increased. **(1)**

2 (a) Reverse the direction of current in the coil **(1)**. Reverse the direction of magnetic flux between the magnets **(1)**.

(b) Any valid example, e.g. washing machine / tumble dryer / airplane propeller / winch **(1)**

3 1. Increase the size of the current flowing in the coil. **(1)**

Answers

2. Increase the magnetic flux density between the poles of the magnet. **(1)**

3. Increase the number of turns on the motor coil. **(1)**

4 The commutator is in contact with the coil **(1)** but it is split into two parts, creating a gap **(1)**. As the motor spins, the contacts touch alternating sides of the commutator **(1)**, causing the current to change direction every half-turn **(1)**.

99 Induced potential

1 (a) Any three from: move the wire faster **(1)**, use a stronger magnetic field **(1)**, use more loops / turns in the wire **(1)**, wind the wire around an iron core **(1)**.

(b) The generator effect **(1)**

2 (a) When a wire or coil moves relative to a magnetic field **(1)** a potential difference is induced **(1)**, resulting in a current being generated. The conductor must cut across the magnetic field for a potential difference to be induced **(1)**.

(b) An induced current generates a magnetic field that opposes the original change. **(1)**

(c) An induced magnetic field may be due to the movement of the conductor **(1)** or a change in the magnetic field **(1)**.

3 (a) Sine wave sketched that goes above and below the x-axis. **(1)**

(b) The current moves between a positive value **(1)** and a negative value **(1)**.

100 Alternators and dynamos

1 Alternating current flows backwards and forwards / alternates in direction **(1)**. Direct current always flows one way / in the same direction **(1)**.

2 D **(1)**

3 (a) Graph A represents current produced by a dynamo. **(1)**

Graph B represents current produced by an alternator. **(1)**

(b) (i) Graph A shows that at every half turn / cycle the current does not change direction. **(1)**

(ii) Graph B shows that at every half turn / cycle the current changes direction. **(1)**

4 An alternator and a dynamo both use the interaction of conductors in magnetic fields **(1)** to produce an electric current **(1)**. As the coil in the alternator rotates, the way it faces is continually changing creating an alternating current through the slip rings **(1)** but, as the coil in a dynamo rotates, the way it faces also changes but, as the contacts also change on the split-ring **(1)** a direct current **(1)** is produced.

101 Loudspeakers

1 A **(1)**

2 A motor converts electrical energy into kinetic energy **(1)**. The loudspeaker does this by converting electrical energy into the movement of the cone **(1)**. This makes use of the magnetic property of electric current **(1)**, which interacts with the magnetic field of a permanent magnet **(1)**, causing a force.

3 The varying force exerted on the cone by the moving coil **(1)** results in movement / vibration of the cone **(1)**. The movement / vibration of the cone pushes the air causing pressure / sound waves **(1)**.

4 (a) A change in frequency of the vibration of the cone will result in a change of pitch in the sound. **(1)**

(b) A change in amplitude of the vibration of the cone will result in a change in volume / loudness of the sound. **(1)**

102 Transformers

1 A step-up transformer is used to increase potential difference and decrease current **(1)**. A step-down transformer is used to decrease the voltage and increase the current **(1)**.

2 When potential difference is increased the current is reduced **(1)**. The lower current **(1)** produces less heating effect **(1)** and therefore less energy is wasted in transmission of electricity **(1)**.

3 $V_p / V_s = n_p / n_s$, $V_p = 230$ V, $V_s = 19$ V, $n_s = 380$ turns; $n_p = V_p n_s / V_s$ **(1)** = (230 V × 380 turns) / 19 V **(1)** = 4600 turns **(1)**

4 (a) This is a step-down transformer **(1)** because there are more turns on the primary coil than on the secondary coil **(1)**.

(b) $n_p = 600$ turns, $n_s = 20$ turns, $V_p = 360$ V, $V_s = (V_p n_s) / n_p$ **(1)** = (360 V × 20 turns) / 600 turns **(1)** = 12 **(1)** V

103 Extended response – Magnetism and electromagnetism

The answer should include some of the following points: **(6)**

- A long straight conductor could be connected to a cell, an ammeter and a small resistor to prevent overheating in the conductor.
- When the current is switched on the direction of the magnetic field generated around a long straight conductor can be found using the right-hand grip rule.
- The right-hand grip rule points the thumb in the direction of conventional current and the direction of the fingers show the direction of the magnetic field.
- A card can be cut halfway through and placed at right angles to the long straight conductor. A plotting compass can be used to show the shape and direction of the magnetic field.
- The shape of the magnetic field around the long straight conductor will be circular/ concentric circles as the current flows through it.
- The strength of the magnetic field depends on the distance from the conductor.
- The concentric magnetic field lines mean that the field becomes weaker with increasing distance.
- The strength of the magnetic field can be increased by increasing the current.

104 The Solar System

1 Sun, planets, dwarf planets, moons (all required for mark) **(1)**

2 A planet orbits the Sun / a star **(1)** but a moon orbits a planet. **(1)**

3 Pluto was re-classified as a dwarf planet **(1)** because more powerful telescopes **(1)** have found many other objects similar to Pluto. **(1)**

4 (a) The Solar System has only one star **(1)** but the Milky Way contains over one billion stars. **(1)**

(b) The Solar System is a simple star system / has orbiting planets / is comparatively small **(1)** but the Milky Way is a spiral galaxy / collection of stars / contains planets **and** stars / is much bigger (accept any valid description of the Milky Way (accept any other valid difference). **(1)**

5 (a) 1. terrestrial / rocky planets **(1)** 2. gas giants. **(1)**

(b) C **(1)**

(c) Saturn is a gas giant whereas the other three planets are terrestrial / rocky planets. **(1)**

(d) The Earth has a solid crust with a separate gaseous atmosphere / atmosphere was produced through geological processes **(1)** whereas the surface of Neptune is gaseous and there seems to be no definite boundary between the 'surface' and the 'atmosphere'. **(1)**

105 The life cycle of stars

1 (nuclear) fusion **(1)**

2 B **(1)**

3 All stars exist for a time as main sequence stars but for massive stars this time is much shorter than for small stars **(1)**. After the main sequence stage, massive stars become red **supergiants** whereas smaller stars become red **giants** **(1)**. After the red supergiant stage, massive stars will explode as supernovae **(1)**, whereas smaller stars collapse to form white dwarfs **(1)**. Massive stars then go on to form either **neutron stars** or, for the most massive stars, **black holes** **(1)**. (Both points needed for final mark.)

4 All of the naturally occurring elements in the periodic table are produced by fusion processes in stars **(1)**. Elements heavier than iron **(1)** are produced in supernova explosions **(1)**. These explosions of massive stars **(1)** then distribute the elements throughout the Universe **(1)**.

106 Satellites and orbits

1 D **(1)**

2 The orbit of a planet is around a star (accept the Sun) **(1)** whereas the orbit of a moon is around a planet **(1)**.

3 (a) Planets, moons and artificial satellites all move in circular **(1)** orbits. (b) A moon orbits at a fixed distance **(1)** from its planet but the orbit of an artificial satellite can be changed by adjusting its speed **(1)** and the radius of its orbit **(1)**.

4 (a) A satellite is held in a circular orbit due to a balance **(1)** between the force provided by gravity **(1)** and its speed **(1)**. (b) A satellite in a stable orbit will move with changing velocity but unchanged speed because the satellite is continually changing direction **(1)** and velocity is a vector **(1)**. (c) If the speed of a satellite changes, the radius of its orbit **(1)** must also change **(1)**, to regain a stable orbit **(1)**.

107 Red-shift

1 D **(1)**

2 A galaxy with a red-shifted spectrum indicates that the galaxy is moving away from the observer **(1)**. The further the black lines are shifted the faster it is moving away **(1)**. As most galaxies are red shifted this would suggest that the Universe is expanding **(1)**.

3 (a) The Big Bang theory. **(1)**

 (b) The observation of supernovae **(1)** in red-shift **(1)**.

 (c) This suggests that the Universe began from one very small region **(1)** that was extremely dense (accept hot) **(1)**.

 (d) (i) Dark matter **(1)** and dark energy **(1)**.

 (ii) The Universe is believed to be expanding even faster than was first thought **(1)**. There is much about the Universe that we do not yet understand **(1)**.

108 Extended response – Space physics

The answer should include some of the following points: **(6)**

- The Big Bang theory states that the Universe started from a very small, hot and dense space in a massive explosion about 13.8 billion years ago.
- The Universe has been expanding and cooling ever since.
- The observed red-shift of light from galaxies provides evidence that the universe is expanding and supports the Big Bang Theory.
- There is an observed increase in the wavelength of light from most distant galaxies.
- The further away the galaxies are, the faster they are moving and the bigger the observed increase in wavelength. This effect is called red-shift.
- The observed red-shift provides evidence that space itself (the Universe) is expanding and supports the Big Bang theory.
- Scientists were able to either repeat the same experiment or test the theory using another method so the Big Bang theory was accepted.
- Since 1998 onwards, observations of supernovae suggest that distant galaxies are receding ever faster so this may lead to a theory that replaces the Big Bang theory.
- If future experiments find a better explanation of the origin of the Universe, then this theory will be adapted or replaced.

109 Timed Test 1

1 (a) (i) Background radiation **(1)**

 (ii) Any two from: radon gas, **(1)** cosmic rays **(1)**, medical uses **(1)**, nuclear industry **(1)**, natural sources/rocks **(1)**

 (iii) Several readings are taken to identify anomalies to gain a reading close to the true value. **(1)**

 (b) (i) *y*-axis labelled 'Corrected count rate (counts / min)' and *x*-axis labelled 'Time in minutes'; **(1)** all points plotted correctly (to within half a square) **(2)** [or six or more points correctly plotted (to within half a square) for 1 mark].

 (ii) Single exponential decay curve of best fit passing through six points **(1)**

 (iii) Half-life = 2 minutes **(1)**

 (c) Alpha radiation is highly ionising because it is the most massive particle so it can easily ionise atoms by knocking off their electrons **(1)**. Beta radiation is moderately ionising; the particles are highly energised although very small so have less chance of knocking electrons off other atoms **(1)**. Gamma radiation is the least ionising because it has no mass but the gamma waves may still ionise other atoms **(1)**.

2 (a) (i) $E_p = m\,g\,h$
 Energy gained = 750 kg × 10 N/kg × 15 m **(1)** = 112 500 **(1)** J

 (ii) Power $P = E_p / t$ = 112 500 J / 20 s **(1)** = 5625 **(1)** W **(1)**

 (b) Any one of: thermal energy store of the motor **(1)**, thermal energy store of the elevator materials **(1)**, thermal energy store of the environment **(1)**, sound energy store **(1)**.

 (c) $v = d / t$ = 15 m / 20 s **(1)** = 0.75 m/s **(1)**
 Kinetic energy, $E_k = \tfrac{1}{2}\,m\,v^2 = \tfrac{1}{2} \times 750$ kg × (0.75 m/s)^2 **(1)** = 210.9 **(1)** J

 (d) This process could be described as wasteful because it causes a rise in temperature in parts of the system so dissipating energy in heating the surroundings. **(1)**

 (e) A rise in temperature in any one of: the lift motor **(1)**, the fabric of the lift **(1)**, the lift cables **(1)**.

3 (a) (i) $P = V\,I$ / power = voltage × current **(1)**

 (ii) $I = P / V$ = 2000 W / 230 V **(1)** = 8.7 A **(1)**

 (iii) Too high a fuse could result in too much current and the risk of a fire **(1)**; too low a fuse would melt and break the circuit each time the kettle was switched on **(1)**.

 (iv) C **(1)**

 (b) 2000 W = 2000 J/s so E = 2000 J **(1)** and $E = Q\,V$, so $Q = E / V$ = 2000 J / 230 **(1)** V = 8.7 **(1)** coulombs / C

4 (a) Energy supplied $E = I\,V\,t$ = 12 A × 12 V × 120 s **(1)** = 17 280 **(1)** joules / J **(1)**

 (b) $\Delta E = m\,c\,\Delta\theta$ so $\Delta\theta = \Delta E / (m\,c)$ = 17 280 J / (2 kg × 385 J/kg °C) **(1)**
 Temperature change = 22.4 **(1)** degrees Celsius / °C **(1)**

 (c) Some thermal energy is dissipated to the environment. **(1)**

 (d) (i) By thermally insulating the block **(1)**

 (ii) Insulator **(1)**

5 (a) (i) $^{231}_{91}\text{Pa} \rightarrow\ ^{227}_{89}\text{Ac} +\ ^{4}_{2}\text{He}$
 i.e. 231, 89, 2 **(1)** (all three needed for mark)

 (ii) Alpha decay **(1)**

 (iii) $^{211}_{87}\text{Fr} \rightarrow\ ^{4}_{2}\text{He} +\ ^{207}_{85}\text{At}$
 i.e. 211, 4, 85 **(1)** (all three needed for mark)

 (iv) Alpha decay **(1)**

 (v) $^{24}_{11}\text{Na} \rightarrow\ ^{24}_{12}\text{Mg} +\ ^{0}_{-1}\text{e}$
 i.e. 12, 0 **(1)** (both needed for mark)

 (vi) Beta decay **(1)**

 (vii) $^{201}_{79}\text{Au} \rightarrow\ ^{201}_{80}\text{Hg} +\ ^{0}_{-1}\text{e}$
 i.e. 79, 201 **(1)** (both needed for mark)

 (viii) Beta decay **(1)**

 (b) (i) Alpha decay causes both the mass and the charge of the nucleus to decrease. **(1)**

 (ii) Beta decay does not cause the mass of the nucleus to change but does cause the charge of the nucleus to increase. **(1)**

6 (a) (i) As the fuel flows through the delivery pipe, electrons can build up due to friction **(1)** because the fuel pipe is made from an insulating material **(1)**.

 (ii) The build-up of static charge on an insulator could result in a spark **(1)**. This would be dangerous because it could ignite the fuel / fumes / vapour **(1)**.

 (b) The further away the galaxies, the faster they are moving **(1)** and the bigger the observed increase in wavelength **(1)**.

 (c) *The answer should include some of the following points:* **(6)**

- A charged object creates an electric field around itself.
- The electric field is strongest close to the charged object.
- The further away from the charged object, the weaker the field.
- A second charged object placed in the field experiences a force.
- The force gets stronger as the distance between the objects decreases.

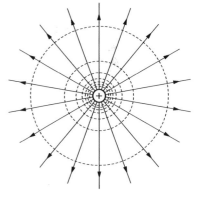

Diagram should include: positive point source, radial field lines, arrows outwards to show direction of field, equipotential lines, closer together nearer the centre.

7 (a) Each lamp would have a potential difference of 0.5 V **(1)** because potential difference is shared in a series circuit **(1)**.

 (b) Voltmeter connected in parallel across one of the lamps. **(1)**

 (c) An additional cell could be added. **(1)**

 (d) The lamps could be connected in a parallel circuit. **(1)**

 (e) (i) The resistance of a thermistor changes with temperature **(1)**. When the temperature rises more electrons are released **(1)**, causing more current to flow and reducing resistance (accept converse) **(1)**.

(ii) Used in a temperature-sensing circuit. **(1)**

8 (a) Solid – closely packed particles in fixed positions within a regular lattice.

Liquid – closely packed particles, irregular arrangement.

Gas – few particles, spread out in random arrangement (all three correct, **(2)** two correct **(1)**)

(b) *The answer should include some of the following points:* **(6)**

- Particles in a solid have a relatively low amount of kinetic energy compared with the other states of matter
  - and remain in molecular bonding with neighbouring atoms.
- Particles in a liquid have higher amounts of kinetic energy than solids
  - but lower amounts of kinetic energy than gases
  - and remain in weak molecular bonding with / can move around neighbouring atoms.
- Particles in a gas have a high amount of kinetic energy
  - and can move independently of other neighbouring atoms.

9 (a) $\rho_1 = m / V = 1.5$ kg / 0.2 m^3 = 7.5 kg/m^3 **(1)**

$\rho_2 = 0.7$ kg / 0.15 m^3 = 4.7 kg/m^3 **(1)**

$\rho_3 = 0.7$ kg / 0.2 m^3 = 3.5 kg/m^3 **(1)**

so block 1 has the highest density **(1)**

(b) (i) The student must make sure that the line of sight from the eye to the meniscus **(1)** is perpendicular to the scale **(1)** (to avoid parallax errors).

(ii) The student must measure the mass of the rock **(1)**. This can be done on a balance or hung from a suspended force meter **(1)**.

10 (a) Thompson's 'plum pudding' model referred to the atom as a positively charged sphere **(1)** with negative charges / electrons distributed throughout the sphere **(1)**. Rutherford's nuclear model of the atom showed the positive charges concentrated in a tiny nucleus **(1)** with the negative charges / electrons orbiting around the outside of the nucleus **(1)**.

(b) *The answer should include some of the following points:* **(6)**

- Alpha particles directed at gold foil
- Most alpha particles pass straight through
  - so most of the atom is empty space.
- A few alpha particles deflected through large angles
- Mass is concentrated at the centre of the atom
  - so the nucleus must be tiny.
- Nucleus is (positively) charged
  - so the negative charges orbit around the dense positive nucleus.

113 Timed Test 2

1 (a) Measure, mark and record the distance to be travelled by the trolley **(1)**, place the trolley at the start / top of the ramp and release **(1)**, time how long it takes to cover the marked distance **(1)** (both distance and timing should be mentioned for the mark). Repeat the experiment to reduce the influence of random errors **(1)**.

(b) Any two from: light gates **(1)**, data logger **(1)**, computer **(1)**

(c) (i) Independent variable: Height **(1)**

(ii) Column 2 and column 3: Distance, Time **(1)** (both needed for mark)

(iii) Speed (m/s) **(1)**

(d) (i) The forces acting opposite to the motion of the trolley / causing the trolley to stop would be friction between the wheels and the surface / floor / between the axles and wheels **(1)**

(ii) Friction could be reduced by (any one of the following): using a smoother surface on the ramp / smoother surface on the wheels / lubricating the axles **(1)**. Air resistance could be reduced by (any one of the following): making the trolley more aerodynamic / reducing the surface area of the front of the trolley **(1)**.

2 (a) $a = (v - u) / t$ **(1)**

(b) Average speed = 5 m/s, so s = 5 m/s × 20 s **(1)**

Distance = 100 m **(1)**

(c) $p = m v$ so momentum = 1000 kg × 10 m/s **(1)** = 10 000 **(1)** kg m/s **(1)**

(d) C **(1)**

3 (a) C **(1)**

(b) (i) Nuclear fusion is the joining of two light nuclei to form a heavier nucleus. **(1)**

(ii) Fusion processes in stars produce all the naturally occurring elements **(1)**. Fusion processes lead to the formation of new elements **(1)**.

(iii) Elements heavier than iron are produced in a supernova **(1)**. The explosion of a massive star (supernova) distributes the elements throughout the Universe **(1)**.

4 (a) (i) time period = 1 / frequency or $T = 1 / f$. **(1)**

(ii) 'T' should be marked between the start and end of an identified single wave cycle. **(1)**

(iii) First find T so T = 0.0001 / 2 (2 waves shown) so T = 0.00005 **(1)** so T = 1 / 0.000 05 s **(1)** = 20 000 **(1)** Hz.

(b) (i) Sonar **(1)**

(ii) The ship emits sound waves, which travel down to the shoal of fish **(1)**. Detectors on the ship receive the echo of the sound waves as they are reflected back from the fish **(1)**. The depth / location of the fish can be found by calculating the time between the sound wave being sent and the echo being detected **(1)**.

5 (a) Reflection **(1)**

(b) (i) *i* marked between incident ray and normal **(1)**, *r* marked between normal and refracted ray (inside the block) **(1)**.

(ii) As the wave passes from a less dense medium (air) to a denser medium (glass) the speed of light slows down **(1)** as the 'leading edge' of the wave front reaches the glass first. The wave then changes direction as the whole wave pivots on this 'leading edge' **(1)**.

(c) Rays drawn – tip of object to mid-lens then down through F_2 **(1)**. Second ray from tip of object through centre of lens to meet first ray **(1)**. Position of image shown **(1)**.

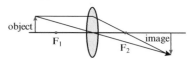

6 (a) A light source that is red-shifted will have a greater wavelength **(1)** and a lower frequency **(1)**.

(b) The further away the galaxies, the faster they are moving **(1)** and the bigger the observed increase in wavelength **(1)**.

(c) Red shift provides evidence of an expanding universe **(1)**, which supports the Big Bang model **(1)**.

(d) CMB radiation / cosmic microwave background radiation **(1)** which is detected from everywhere **(1)**.

(e) *The answer should include some of the following points:* **(6)**

- Scientists think that galaxies would rotate much faster if the stars they can see and detect were the only matter in galaxies. **(1)**
- Scientists have therefore proposed that there must be matter that cannot be detected or seen. **(1)**
- The concepts of dark mass **(1)** and dark energy **(1)** are not yet understood, **(1)** but one day may account for the 'missing matter' **(1)**.

7 (a) Planets and dwarf planets orbit round the Sun **(1)** but moons orbit round planets **(1)**.

(b) Jupiter is much larger than Mars / Jupiter is the largest planet in the Solar System **(1)**. Jupiter has many more moons than Mars / Jupiter has over 50 confirmed moons whereas Mars has 2 **(1)**.

(c) The circular path means a continual change of direction **(1)** and direction is a component of velocity **(1)**.

(d) *The answer should include some of the following points:* **(6)**

- At the start of a star's life cycle, the dust and gas / nebula are drawn together by gravity. **(1)**
- This causes fusion reactions. **(1)**
- The fusion reactions lead to an equilibrium between:
  - the gravitational collapse of a star **(1)** and
  - the expansion of a star due to fusion energy **(1)**.
- As the star loses mass, it expands to become a red giant. **(1)**
- The red giant then cools and collapses to become a white dwarf **(1)**.

8 (a) (i) The direction of the magnetic field will be clockwise **(1)**, as the current flows from positive to negative **(1)** (see diagram).

(ii) The right-hand rule **(1)**

(b) (i)

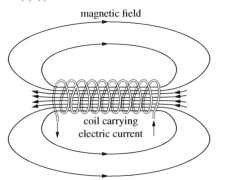

magnetic field

coil carrying electric current

Correct shape **(1)** lines of flux very close inside **(1)** lines of flux further apart outside, **(1)** pattern indicated on both sides of electromagnet. **(1)**

(ii) Any valid use, e.g. doorbell, speaker, electromagnet **(1)**

(c) $F = B\,I\,l = 0.5\text{ T} \times 3\text{ A} \times 0.75\text{ m}$ **(1)** $= 1.1$ **(1)** N **(1)**

9 (a) The step-up transformer increases the potential difference leaving the power station so that the current is reduced **(1)**. This reduces the amount of energy transferred to the thermal store / wasted during the transmission of electricity through the National Grid **(1)**.

(b) $V_p / V_s = n_p / n_s$ so $V_s = (n_s\,V_p) / n_p$ **(1)** $= (4500 \times 250\text{ V}) / 150$ **(1)**

Potential difference = 7500 **(1)** V

(c) The high voltages that are necessary for efficient transmission via the National Grid are too dangerous **(1)** so a step-down transformer is used to reduce the voltage **(1)**.

(d) (i) $V_p\,I_p = V_s\,I_s$ so $I_s = (V_p\,I_p) / V_s$ **(1)** $= (4600\text{ V} \times 0.5\text{ A}) / 230\text{ V}$ **(1)**

Current = 100 **(1)** A

(ii) Assume that the transformer is 100% efficient. **(1)**

10 (a) The students have not unloaded the spring each time to check that the deformation is elastic. **(1)**

(b) The students should use the same weights to get more reliable results to show a proportional relationship **(1)**; they should unload the spring after measuring each extension to check that the deformation is still elastic **(1)**.

(c) 0.1:36, 0.2:40, 0.3:44, 0.5:52 (all four correct **(2)**; three correct **(1)**)

(d) In linear elastic deformation, the extension (of a spring) is directly proportional to the force added **(1)**. The graph is a straight line which shows a directly proportional relationship **(1)**. The spring is loaded with weights of 0.1 N at a time **(1)** and the spring extends by 4 mm with each 0.1 N **(1)**.

Physics Equations Sheet

1	pressure due to a column of liquid = height of column × density of liquid × gravitational field strength (g)	$p = h \rho g$
2	(final velocity)^2 − (initial velocity)^2 = 2 × acceleration × distance	$v^2 - u^2 = 2\,a\,s$
3	force $= \dfrac{\text{change in momentum}}{\text{time taken}}$	$F = \dfrac{m\,\Delta v}{\Delta t}$
4	elastic potential energy = 0.5 × spring constant × (extension)^2	$E_e = \dfrac{1}{2}\,k\,e^2$
5	change in thermal energy = mass × specific heat capacity × temperature change	$\Delta E = m\,c\,\Delta\theta$
6	period $= \dfrac{1}{\text{frequency}}$	
7	magnification $= \dfrac{\text{image height}}{\text{object height}}$	
8	force on a conductor (at right angles to a magnetic field) carrying a current = magnetic flux density × current × length	$F = B\,I\,l$
9	thermal energy for a change of state = mass × specific latent heat	$E = m\,L$
10	$\dfrac{\text{potential difference across primary coil}}{\text{potential difference across secondary coil}} = \dfrac{\text{number of turns in primary coil}}{\text{number of turns in secondary coil}}$	$V_p / V_s = n_p / n_s$
11	potential difference across primary coil × current in primary coil = potential difference across secondary coil × current in secondary coil	$V_s\,I_s = V_p\,I_p$
12	For gases: pressure × volume = constant	$p\,V = \text{constant}$

Your own notes

Your own notes

Your own notes

Your own notes

Your own notes

Your own notes

Your own notes

Published by Pearson Education Limited, 80 Strand, London, WC2R 0RL.

www.pearsonschoolsandfecolleges.co.uk

Text and illustrations © Pearson Education Limited 2017
Typeset, illustrated and produced by Phoenix Photosetting
Cover illustration by Miriam Sturdee

First published 2017

20 19 18 17
10 9 8 7 6 5 4 3 2 1

British Library Cataloguing in Publication Data
A catalogue record for this book is available from the British Library

ISBN 978 1 292 13150 4

Printed in Slovakia by Neografia

Acknowledgements
Content by Peter Ellis is included in this book.

Page 9, Patterns of world energy use, graph includes data from BP Statistical Review of World Energy 2012 www.bp.com/statisticalreview; page 74, Highway Code stopping distances: Contains public sector information licensed under the Open Government Licence v1.0.

Note from the publisher
Pearson has robust editorial processes, including answer and fact checks, to ensure the accuracy of the content in this publication, and every effort is made to ensure this publication is free of errors. We are, however, only human, and occasionally errors do occur. Pearson is not liable for any misunderstandings that arise as a result of errors in this publication, but it is our priority to ensure that the content is accurate. If you spot an error, please do contact us at resourcescorrections@pearson.com so we can make sure it is corrected.

Printed in Great Britain
by Amazon

Extended response questions

Answers to 6-mark questions are indicated with a star (*).

In your exam, your answers to 6-mark questions will be marked on how well you present and organise your response, not just on the scientific content. Your responses should contain most or all of the points given in the answers below, but you should also make sure that you show how the points link to each other, and structure your response in a clear and logical way.

1 Energy stores and systems

1 A – gravitational, C – chemical, D – kinetic **(1)** (all three needed for mark)

2 (a) A closed system is an isolated system where no energy flows in or out of the system. **(1)**

(b) The total energy in a closed system is the same after the transfer as it was before the transfer **(1)** because energy cannot be created or destroyed **(1)**.

3 (a) Store of chemical energy **(1)**.

(b) Energy transfer by electrical current **(1)**.

(c) Energy transfer to the surroundings by sound waves and by heating **(1)** (both needed for mark).

4 The total energy available initially in this closed system, in the gravitational potential store, is 250 J **(1)**. As the basket reaches the ground, the gravitational potential store will become 0 J **(1)**, because it has been transferred to a total of 250 J of useful kinetic energy **(1)** and wasted thermal energy **(1)**.

2 Changes in energy

1 D **(1)**

2 Kinetic energy $E_k = \frac{1}{2} m v^2$ so $E_k = \frac{1}{2} \times 70 \times 6^2$ **(1)** = 1260 **(1)** J / joules **(1)**

3 Convert units: 15 cm = 0.15 m **(1)**

Energy transferred $E = \frac{1}{2} k x^2 = \frac{1}{2} \times 200 \text{ N/m} \times (0.15 \text{ m})^2$ **(1)** = 2.25 **(1)** J

4 Change weight to mass first, using $m = F / g$ so 600 / 10 = 60 kg **(1)**

Gravitational potential energy (GPE) = 60 kg $\times$ 10 N/kg $\times$ 0.7 m **(1)** = 420 **(1)** J

3 Energy changes in systems

1 Specific heat capacity = change in thermal energy ÷ (mass × change in temperature) or $(c = \Delta E \div (m \times \Delta T)$ **(1)**

2 $\Delta E = m c \Delta\theta$ **(1)** = 0.8 kg $\times$ 4200 J / kg °C $\times$ 50 °C **(1)** = 168 000 **(1)** J

3 $\Delta\theta = \Delta E / (m c)$ so $\Delta\theta$ = 20 000 J / (1.2 kg × 385 J/kg °C) **(1)** = 43 **(1)** so

Change in temperature of the copper = 43 °C **(1)**

4 $E_{in} = P t$ = 30 W $\times$ 540 s **(1)** = 16 200 J **(1)**

$\Delta E_{in} = \Delta E_{out}$ so

Specific heat capacity c = 16 200 J / (0.8 kg $\times$ 25 °C) **(1)** = 16 200/20 = 810 **(1)** J / kg °C

4 Specific heat capacity

1 (a) the amount of energy required to raise the temperature of 1 kg of material by 1 K (or 1 °C) **(1)**

(b) Energy supplied, mass and change in temperature. **(1)**

2 (a) Place a beaker on a balance, zero the balance and add a measured mass of water **(1)**. Take a start reading of the temperature **(1)**. Place the electrical heater into the water and switch on **(1)**. Take a temperature reading every 30 seconds **(1)** until the water reaches the required temperature **(1)**.

(b) Measure the current supplied, the potential difference across the heater and the time for which the current is switched on **(1)**. Use these values to calculate the thermal energy supplied using the equation $E = V \times I \times t$ **(1)**.

(c) Add insulation around the beaker **(1)** so less thermal energy is transferred to the surroundings and a more accurate value for the specific heat capacity of the water may be obtained. **(1)**

3 Plot a graph of temperature against time **(1)**. The changes of state are shown when the graph is horizontal (the temperature is not increasing) **(1)**.

5 Power

1 A **(1)**

2 Energy transferred = 15 000 J, time taken = 20 s

$P = E / t$

so power P = 15 000 J / 20 s **(1)** = 750 **(1)** W

3 (a) $E_g = m g h$ so E_g = 60 × 10 × (0.08 × 20) = 960 **(1)** joules / J **(1)**

(b) $P = E / t$ so P = 960 / 12 = 80 watts / W **(1)** (allow value for energy calculated in (a))

4 (a) For 3 W motor: t = 360 J ÷ 3 W **(1)** = 180 **(1)** s.

(b) For 5 W motor: t = 360 J ÷ 5 W **(1)** = 72 **(1)** s

6 Energy transfers and efficiency

1 (a) Concrete **(1)**

(b) Low relative thermal conductivity means that a material will have a slow **(1)** rate of transfer of thermal energy. **(1)**

2 (a) Thicker walls provide more material for the thermal energy **(1)** to travel through from the inside to outside, so the rate of thermal energy **(1)** loss is less, keeping the houses warmer.

(b) Thicker walls provide more material for the thermal energy **(1)** to travel through from the outside to inside, so the rate of thermal energy transfer **(1)** is less, keeping the house cool.

3 (a) The useful energy transferred to the box = 100 J; total energy used by the motor = 400 J; efficiency = 100 J / 400 J **(1)** = 0.25. **(1)** (accept × 100 = 25%.)

(b) Using lubrication between moving surfaces will reduce friction **(1)** and therefore reduce wasted thermal energy **(1)**.

4 A **(1)**

7 Thermal insulation

1 (a) Any **five** of the following points: select a minimum of four beakers, one to remain unwrapped as a control **(1)**, wrap three or more beakers with the same mass / thickness of insulating material **(1)**, provide insulating bases and lids (with a hole for the thermometer) **(1)**, add the same amount of boiling or very hot water to the beakers **(1)**, place thermometers in each beaker and record the temperature of the water at the highest point **(1)**, start stop watch **(1)**, and if recording a temperature curve, take temperature reading every minute (or other time increment) **(1)**.

(b) Independent variable – type of insulation selected **(1)**; dependent variable – temperature **(1)**

(c) Control variables: any four from – mass / thickness of insulating material, volume of water, starting temperature of the water, size of beaker, material of beaker, time of experiment. (For four **(1)**, for three **(1)**)

2 Any one of the following: hot water in eye can cause damage – always wear eye goggles **(1)**; hot water can cause scalds – place kettle close to beakers **(1)**; control beaker, and others, can cause burns – do not touch **(1)**; spilt water can cause slippage – report and wipe up **(1)**; and glass thermometers can be broken and cause cuts – handle with care (same if using glass beakers) **(1)**.

3 A digital thermometer can improve accuracy in reading the temperature. **(1)**

4 The greater the thickness of insulating material, the slower the rate at which the hot water cools **(1)**. The lower the thermal conductivity of the insulating material, the lower the rate at which the hot water will cool **(1)**.

8 Energy resources

1 (a) A hydroelectric power station is a reliable producer of electricity because it uses the gravitational potential energy of water which can be stored until it is needed **(1)**. As long as there is no prolonged drought / lack of rain the supply should be constant **(1)**.

(b) Any **one** of the following: hydroelectric power stations have to be built in mountainous areas / high up (compared with supply areas, so that the gravitational

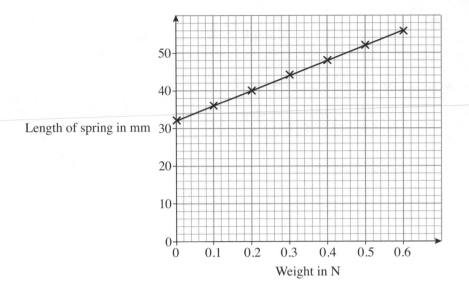

(d) Explain how the graph illustrates the relationship between the length of the spring and the weight added. You should refer to the graph in your answer.   **(4 marks)**

potential energy can be captured) (1); the UK has very few mountainous areas like this (1); and limited to areas such as North Wales and Scottish Highlands (1).

2 (a) When carbon dioxide is released into the atmosphere it contributes to the greenhouse effect / build-up of CO_2 (1) which is believed to contribute to global warming (1).

(b) Sulfur dioxide and nitrogen oxides have been found to dissolve in the water droplets in rainclouds, increasing their acidity (1); this can kill plants and forests / lakes or dissolve the surfaces of historical limestone buildings (1).

(c) Coal mines / oil / gas wells create environmental scars on the landscape (1). Vehicles used to transport fossil fuels add to environmental pollution (1). Alternative answers may include: Accidents in the extraction of oil from deep sea reserves can result in sea pollution (1). New methods of extraction may impact on previously unused areas of the environment e.g. fracking (1). Large transport networks may be needed to transport fuels (1).

9 Patterns of energy use

1 (a)
1. After 1900, the world's energy demand rose / increased (1) as the population grew.
2. There was development in industry / demand in energy supply. (1)
3. The rise of power stations using fossil fuels added to demand. (1)

(b) (i) Coal, oil and natural gas (1) (all three needed)

(ii) Any two from: population has increased so domestic energy use has increased (1); industry has grown, requiring more energy (1); vehicle use and travel have grown, requiring more energy (1) (and any other valid reason)

(iii) Nuclear research only began from the 1940s onwards. (1)

(iv) Hydroelectric. (1)

2 As the population continues to rise the demand for energy will also continue to rise (1). Current trends show that the use of fossil fuels is the major contributor to the world's energy resources (1). These are running out and no other energy resource has, so far, taken their place (1). This could lead to a large gap between demand and supply (1). (Any other valid reason)

10 Extended response – Energy

The answer should include some of the following points: (6)
- Refer to the change in gravitational potential energy (E_p) as the swing seat is pulled back/raised higher.
- Before release, the E_p is at maximum/kinetic energy (E_k) of the swing is at a minimum.
- When the swing is released, the E_p store falls and the E_k store increases.
- E_k is at a maximum at the mid-point, E_p is at a minimum.

- The system is not 100% efficient; some energy is dissipated to the environment.
- Friction due to air resistance and/or at the pivot results in the transfer of thermal energy to the surroundings/environment.
- Damping, due to friction, will result in the E_k being transferred to the thermal energy store of the swing and hence to the environment.
- Eventually all the E_p will have been dissipated to the surroundings/environment (so is no longer useful).

11 Circuit symbols

1 C (1)
2 (a) 1. Resistor (1)
    2. Fuse (1)
    3. Variable resistor (1)
  (b) Fuse (1)
3

Component	Symbol	Purpose
ammeter	(A)	measures electric current (1)
fixed resistor (1)	⊏⊐	provides a fixed resistance to the flow of current
diode (1)	▷⊢	allows the current to flow one way only
switch (1)	⌐o⟋ o⌐ or ⌐o o⌐	allows the current to be switched on / off

(Each correctly completed row gains 1 mark.)

4 Diagram showing series circuit diagram with battery / power supply (1). Resistor (1) with ammeter in series (1). Voltmeter connected in parallel across the resistor (1).

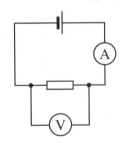

12 Electrical charge and current

1 (a) An electric current is the rate (1) of flow of charge (electrons in a metal) (1).
  (b) Charge $Q = I\,t = 4\,A \times 8\,s$ (1) = 32 (1) coulombs / C (1)
2 (a) (i) The current is the same in all parts of a series circuit so the readings on ammeter 1 and ammeter 3 will be the same as that shown for ammeter 2. (1)
    (ii) Add another cell / increase the energy supplied. (1)
  (b) Cell (1)
3 (a) Any series circuit diagram with a component (e.g. lamp) (1) and an ammeter (1).
  (b) stopwatch / timer (1)

13 Current, resistance and pd

1 D (1)
2 Ohm's law states: The current flowing through a resistor (1) at constant temperature is directly proportional to the potential difference across the resistor. (1)
3 (a) Resistance $R = V / I = 12\,V / 0.20\,A$ (1) = 60 Ω (1)
  (b) Current $I = V / R$ so $I = 22\,V / 55\,Ω$ (1) = 0.40 A (1)
4 (a) Line A – straight line through origin (1)
    Line B – straight line through origin – different gradient (1)
  (b) Line with smaller gradient (1)

14 Investigating resistance

1 B (1)
2 $R = V / I$ so $R = 90 / 1.5$ (1) = 60 (1) Ω.
3 All circuit symbols correct (2 cells, 2 lamps, 1 ammeter, 1 voltmeter, wire) (1). Ammeter connected in series (1). Voltmeter connected in parallel with one lamp (1).

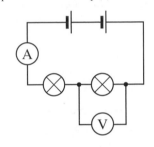

4 A Fixed resistor: The temperature remains constant so the resistance remains constant, as shown by the straight line on the graph. (1) (both needed)
  B Filament lamp: As potential difference increases, the filament gets hot, so resistance increases, as shown by the curved line on the graph. (1) (both needed)
  C Diode: The current flows in only one direction and the resistance is constant, as shown by the straight line on the graph. (1) (both needed)

15 Resistors

1 (a) As the potential difference increases, the current increases (1) in a linear / proportional relationship (1).
  (b) As the potential difference increases the current increases (1) but the gradient of the line gets less steep / shallower, or increase in current becomes smaller as the potential difference continues to increase (1).

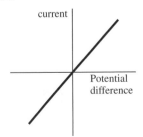

2 (a) Fixed resistor: as graph A in Q1 **(1)**

   Filament lamp: as graph B in Q1 **(1)**

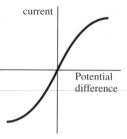

current | Potential difference

(b) The different shaped graphs are because the fixed resistor is ohmic / obeys Ohm's law, so the current and potential difference have a proportional relationship **(1)**; the filament lamp does not obey Ohm's law so the relationship between current and potential difference is not proportional / current begins to level off as potential difference increases **(1)**.

3 Data can be collected using an ammeter to measure current **(1)** and a voltmeter to measure potential difference **(1)**. A wire should be included and a fixed resistor **(1)** to prevent overheating. A range of potential difference **(1)** measurements should be made so that resistance can be calculated using the equation $R = V / I$ **(1)**.

16 LDRs and thermistors

1

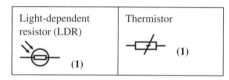

Light-dependent resistor (LDR)	Thermistor
(1)	**(1)**

2 (a) The resistance goes down as the light becomes more intense (brighter) (more current flows). **(1)** (b) The resistance goes down as the temperature goes up (more current flows). **(1)**

3 The thermistor reacts to rise in temperature **(1)** in the engine. Above a certain temperature, it allows current **(1)** in the circuit to flow to a fan, which cools **(1)** the engine.

4 In the light, the resistance of the LDR is low **(1)** so less current flows through the lamp, turning the light off (in the light, more current will flow through the LDR instead of through the lamp) **(1)**.

17 Investigating *I*–*V* characteristics

1 (a) Ammeter connected in series. **(1)** Voltmeter connected in parallel across the component to be tested. **(1)**

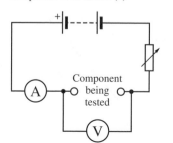

A

Component being tested

V

(b) (i) Potential difference (*V*) **(1)**

(ii) Current (*I*) **(1)** (can be in reverse order)

(c) The terminal connections should be reverse to obtain negative values. **(1)**

(d) *y*-axis: current (*I*), *x*- axis: potential difference (*V*) **(1)** (both needed)

2 (a) wire / ohmic resistor **(1)**

(b) filament lamp **(1)**

(c) diode **(1)**

3 Ohm's law: resistance = potential difference / current **(1)**

R (Ω or ohms) = V (V or volts) / I (A or amps/ amperes) **(1)** (1 mark for all three symbols, 1 mark for all three units)

4 Resistors can become hot and cause burns or fire. **(1)**

18 Series and parallel circuits

1 (a) In a series circuit the current flowing through each component is the same **(1)**. In a parallel circuit, the current is shared between the components **(1)**.

(b) Series: A_2 = 3 A, A_3 = 3 A **(1)** (both needed)

Parallel: A_2 = 1 A, A_3 = 1 A, A_4 = 1 A **(1)** (all three needed)

2 (a) In a series circuit, the total potential difference supplied is shared between the components **(1)**. In a parallel circuit, the potential difference across each component is the same as the potential difference supplied **(1)**.

(b) Series: V_2 = 3 V, V_3 = 3 V, V_4 = 3 V **(1)** (all three needed)

Parallel: V_2 = 9 V, V_3 = 9 V, V_4 = 9 V **(1)** (all three needed)

3 The total resistance of two or more resistors arranged in series is equal to **(1)** the sum of the resistance of each component **(1)**. The total resistance of two or more resistors arranged in parallel is less than **(1)** the resistance of the smallest individual resistor **(1)**.

19 ac and dc

1 (a) Direct potential difference is constant **(1)** and the current flows in the same direction **(1)**.

(b) Alternating potential difference is changeable **(1)** and the current constantly changes direction **(1)**.

2 (a) $P = E / t$ so $E = P t$ = 2000 × (15 × 60) s **(1)** = 1 800 000 J **(1)**

(b) E = 10 × (6 × 60 × 60) s **(1)** = 216 000 J **(1)**

3 There should be one horizontal line anywhere on the screen. **(1)**

4 (a) Toaster: $P = E / t$ = 120 000 J / 60 s **(1)** = 2000 W **(1)**

(b) Kettle: P = 252 000 J / 120 s **(1)** = 2100 W **(1)**

Therefore the kettle has the higher power rating. **(1)**

20 Mains electricity

1 Earth wire (green and yellow), **(1)** live wire (brown), **(1)** neutral wire (blue), **(1)** fuse **(1)**

2 (a) Mains electricity is delivered through an alternating **(1)** current.

(b) The potential difference between the live wire and neutral wire is about 230 **(1)** V. The neutral wire is at, or close to, 0 **(1)** V.

(c) The earth wire is at 0 V **(1)** and only carries a current if there is a fault. **(1)**

(d) In the UK, the domestic electricity supply has a frequency of 50 **(1)** Hz.

3 When a large current enters the live wire **(1)**, this produces thermal energy **(1)**, which melts the wire in the fuse **(1)** and the circuit is then broken **(1)**.

4 The earth wire is connected to the metal casing **(1)**. If the live wire becomes loose and touches anything metallic, the user is protected because the current passes out through the earth wire **(1)** rather than through the user **(1)**.

21 Electrical power

1 $P = I V$ = 5 A × 230 V **(1)** Power = 1150 **(1)** W

2 (a) $P = I V$ so $I = P / V$ **(1)** = 3 W / 6 V **(1)** Current = 0.5 **(1)** A

(b) $P = I^2 R$ **(1)** = (0.5 A)^2 × 240 Ω **(1)** Power = 60 **(1)** W

3 D **(1)**

4 Find current: $I = V / R$ so I = 80 V / 8 Ω = 10 A **(1)**

use either: power $P = I V$ so P = 10 A × 80 V **(1)** = 800 **(1)** W

or $P = I^2 R$ so P = 100 A × 8 Ω **(1)** = 800 **(1)** W

22 Electrical energy

1 $E = Q V$ **(1)** = 30 C × 9 V **(1)** = 270 J **(1)**

2 (a) $Q = I t$ so Q = 0.2 A × 4 V = 0.8, so charge flow = 0.8 **(1)** C.

(b) $E = Q V$ = 0.8 C × 30 s **(1)** = 24 so energy transferred = 24 **(1)** joules / J **(1)**

3 (a) The power of a circuit device is a measure of the rate of energy transfer / how fast the energy is transferred by the device. **(1)**

(b) Energy transferred is the amount of power over a given time **(1)** and the power of a device is the product of the current passing through it and the potential difference across it **(1)**.

4 (a) $Q = E / V$ = 600 J / 20 V **(1)** = 30 C **(1)**

(b) $Q = I t$, so $t = Q / I$ = 30 C / 0.15 A **(1)** = 200 **(1)** s

23 The National Grid

1 (a) As the voltage is increased so the current goes down **(1)**, so this reduces the heating effect due to resistance **(1)** and means that less energy is wasted in transmission **(1)**.

(b) Wasting less energy in transmission means that more energy is transferred to where it is needed **(1)**, making the National Grid an efficient way to transmit energy **(1)**.

(c) The voltages are high enough to kill you if you touch or come into contact with a transmission line. **(1)**

2   $P = I V = 20\ 000\ \text{A} \times 25\ 000\ \text{V}$ **(1)** $=$
500 000 kW **(1)**

3   Any two from: step-up transformers increase
voltage and so lower the current, reducing the
heating losses **(1)**. Wires are thermally insulated
(1). Wires of low resistance are used **(1)**.

4   Step-up transformers are used to increase the
voltage **(1)** as it leaves the power station for
transmission through the National Grid **(1)**.
Near homes, step-down transformers are used
to reduce the voltage **(1)** to make it safer for
use in houses **(1)**.

24   Static electricity

1   The student transfers a charge on to the balloon
by transferring electrons from the jumper to the
balloon **(1)**. The negative charges on the balloon
repel the electrons in the wall **(1)**, inducing a
positive charge on the wall **(1)** which attracts the
negatively charged balloon **(1)**.

2   B **(1)**

3   (a) Friction transfers electrons to another
material. **(1)**

(b) Friction transfers electrons from another
material. **(1)**

4   Insulators do not allow electrons to flow
(1). Instead the electrons either collect on
the insulator (building up a charge) **(1)** or
are knocked off (leaving a positive charge)
(1). Conductors allow electrons to flow **(1)**.
Mutually repelling electrons will then flow
away and this dissipates any charge build-up
(1).

25   Electric fields

1   An electric field is a region in space **(1)** where
a charged particle may experience a force. **(1)**

2   High density lines of flux **(1)**, radial field **(1)**,
arrows pointing outwards **(1)**.

radial field

3   The electric field is strongest close to the
charged object **(1)**. The further away from the
charged object, the weaker the field **(1)**.

4   (a) (i) outwards **(1)**

(ii) outwards **(1)**

(iii) inwards **(1)**

(b) In an electric field, electrically charged
particles experience a force **(1)** that
accelerates the particles **(1)**.

5   The student is right **(1)** because the
electrically charged insulator will either be
negative or positive **(1)** (both needed) due to
gaining or losing electrons **(1)**. The charged
insulator then becomes a point source
creating an electric field **(1)**.

26   Extended response – Electricity

*The answer should include some of the
following points:* **(6)**

● The thermistor can be connected in series
with an ammeter to measure current with a

voltmeter connected in parallel across it to
measure potential difference.

● Ohm's law can be referred to in calculating
the resistance.

● When the temperature is low the resistance
of the thermistor will be high, allowing only
a small current to flow.

● When the temperature is high the resistance
of the thermistor will be low, allowing a
larger current to flow.

● The light-dependent resistor can be connected
in series with an ammeter to measure current
with a voltmeter connected in parallel across
it to measure potential difference.

● When light levels are low (dark) the
resistance of the light-dependent resistor
will be high, allowing only a small current
to flow.

● When light levels are high (bright) the
resistance of the light-dependent resistor will
be low, allowing a larger current to flow.

● Thermistors can be used in fire alarms as a
temperature sensor to switch on an alarm.

● Light-dependent resistors can be used in
security systems as a light sensor to switch
on a light.

27   Density

1   (a) 1. Solid

2. Liquid

3. Gas **(1)**

(b) In a solid, mass per unit volume is higher
than for a liquid or a gas because the
particles are very close together **(1)**. In
a liquid, mass per unit volume is lower
than that for a solid, because the particles
are further apart, but higher than that
for a gas because the particles are closer
together **(1)**. In a gas, mass per unit
volume is low because the particles are
furthest apart **(1)**.

(c) 1. In a solid, high mass / number of
particles per unit volume means that
density is high **(1)**.

2. In a liquid, lower mass / number of
particles per unit volume (than in a
solid) means that density is lower
(than in a solid) **(1)**.

3. In a gas, very low mass / number of
particles per unit volume means that
density is very low **(1)**.

2   B **(1)**

3   Volume = 10 cm × 25 cm × 15 cm =
3750 cm³ **(1)**

$\rho = m \div V$

so $m = \rho V = 3$ g/cm³ × 3750 cm³ **(1)** =
11 250 g **(1)** = 11.25 kg **(1)**

28   Investigating density

1   (a) mass **(1)**

(b) electronic balance **(1)**

2   (a) For regularly shaped solids: any one of
the following methods:

1. Volume can be directly measured
using Vernier callipers / ruler to
measure the length, width and height
of the object **(1)**. The measurements
are then multiplied together to find
the volume, e.g. 3 cm × 3 cm × 3 cm
to give the volume of a 3-cm cube **(1)**.

2. If the object is a regular shape, e.g.
cube, cylinder, sphere, prism, the
appropriate mathematical expression
can be used **(1)**, e.g. use $4 / 3\ \pi\ r^3$ to
find the volume of a sphere **(1)**.

3. If the density and mass are already
known the volume can be calculated
by using the equation **(1)**, i.e. volume
= mass / density **(1)**.

(b) For an irregular solid: any one of the
following methods:

1. Pour water into a measuring cylinder
to a specific level and record the
level **(1)**. Add the object to the water
and record the new water level. The
difference between the new water
level and the original level will be the
object's volume **(1)**.

2. Use a Eureka can by filling it with
water until the water runs out from
the spout **(1)**. When no more water
runs out, carefully place the irregular
solid into the can and measure the
volume of water displaced through
the spout by collecting the water in a
measuring cylinder **(1)**.

3   (a) Place a measuring cylinder on a balance
and then zero the scales with no liquid in
the measuring cylinder **(1)**. Add the liquid
and measure the level **(1)**. Record the
mass of the liquid (in g) from the balance
and the volume (in cm³) by reading from
the level in the measuring cylinder **(1)**.

(b) Take the value at the bottom of the meniscus
(1) making sure that the reading is made at
'eye level', to avoid a 'parallax error' **(1)**.

(c) Density = mass / volume so 121 g ÷
205 cm³ **(1)** = 0.59 **(1)** g/cm³

29   Changes of state

1   In a liquid there are some intermolecular
forces between particles as they move round
each other **(1)**. In a gas there are almost no
intermolecular forces as the particles are far
apart **(1)**.

2   Particles have different amounts in the kinetic
energy store **(1)** and experience different
intermolecular forces **(1)**.

3   B **(1)**

4   At boiling point the liquid changes state **(1)**
so the energy applied after boiling point is
reached goes into breaking bonds **(1)** between
the liquid particles. The particles gain more
energy and become a gas **(1)**.

5   The kinetic energy **(1)** of the particles
decreases **(1)** as the ice continues to lose
energy to the surroundings; this is measured
as a fall in temperature **(1)**.

30   Internal energy

1   B **(1)**

2   At boiling point / latent heat of vaporisation
there will be a change / increase to the
potential energy of the particles **(1)** but the
kinetic energy of the particles will not change
(1).

3   (a) When temperature rises due to heating,
internal energy increases **(1)** because the
kinetic energy of the particles increases
(1).

(b) When temperature does not rise, due to heating, internal energy increases (1) because the potential energy of the particles increases (1).

4   When the water vapour condenses into liquid water there will be no change in the kinetic energy (1) of the water particles so the temperature does not change (1) but there will be a change in the potential energy (1) of the water particles as they move from a gas state to a liquid state.

31   Specific latent heat

1   Specific latent heat is the energy that must be transferred to change 1 kg of a material from one state of matter to another. (1)

2   C (1)

3   $E = m\,L = 25\text{ kg} \times 336\,000$ J/kg (1) = $8\,400\,000$ (1) J

4   (a) Melting – B (1)

   (b) Boiling – D (1)

   (c) Specific latent heat of fusion – B (1)

   (d) Specific latent heat of vaporisation – D (1)

   (e) The energy being transferred to the material is breaking bonds (1); as a result, the material undergoes a phase change (1).

5   $E = m\,L = 36\text{ kg} \times 2260$ kJ/kg (1) = $81\,360$ (1) kJ

32   Particle motion in gases

1   Temperature is a measurement of the average kinetic energy of the particles in a material. (1)

2   $273\text{K} \rightarrow 0\,°\text{C}$ (1)

   $255\text{K} \rightarrow -18\,°\text{C}$ (1)

   $373\text{K} \rightarrow 100\,°\text{C}$ (1)

3   At a constant volume, the pressure and temperature of a gas are directly proportional. (1)

4   (a) As the temperature increases the particles will move faster (1) because they gain more energy (1).

   (b) As the particles are moving faster they will collide with the container walls more often (1), therefore increasing the pressure (1).

   (c) It increases (1)

33   Pressure in gases

1   When particles of a gas collide (1) with a surface they exert a force at right angles to the surface (1) resulting in pressure (1).

2   C (1)

3   $P_1 V_1 = P_2 V_2$ so 100 kPa $\times 230$ cm^3 = 280 kPa $\times V_2$. (1) $V_2 = 100$ kPa $\times 230$ cm^3 / 280 kPa (1) = 82.1 (1) cm^3

4   $P_1 = P_2 V_2 / V_1$ (1) = 640 litres $\times 100$ kPa / 8 litres (1) = 8000 kPa (or 8 MPa or 8×10^6 Pa) (1)

34   Extended response – Particle model

The answer should include some of the following points: (6)

- Solid, liquid and gas states of matter have increasing kinetic energy of particles.
- Thermal energy input or output will result in changes to the thermal energy store of the system and will result in changes of state or a change in temperature.
- Changes in states of matter are reversible because the material recovers its original properties if the change is reversed.
- Thermal energy input does not always result in a temperature rise if the energy is used to make or break bonds between particles / result in a change of state.
- Latent heat is the amount of heat / thermal energy required by a substance to undergo a change of state.
- The thermal energy required to change from solid / ice to water (accept converse) is called the latent heat of fusion and is calculated using Q_f = ml.
- The thermal energy required to change from liquid to gas / water to steam (accept converse) is called the latent heat of vaporisation and is calculated using Q_v = ml.

35   The structure of the atom

1   (a) Protons – labelled in the nucleus (+ charge) (1)

   (b) Neutrons – labelled in nucleus (0 charge) (1)

   (c) Electrons – labelled as orbiting (- charge) (1)

2   (a) The number of positively charged protons (1) in the nucleus is equal to the number of negatively charged electrons (1) orbiting the nucleus.

   (b) The atom will become a positively charged ion / charge of +1. (1)

3   Size of an atom: 10^{-10} m (1)

   Size of a nucleus: 10^{-15} m (1)

4   When an electron absorbs electromagnetic radiation (1) it will move to a higher energy level (1). When the electron moves back from a higher energy level to a lower energy level (1) it will emit electromagnetic radiation (1).

36   Atoms, isotopes and ions

1   (a) The name given to particles in the nucleus. (1)

   (b) The number of protons in the nucleus. (1)

   (c) The total number of protons and neutrons in the nucleus. (1)

2   C (1)

3   Isotopes will be neutral because the number of positively charged protons (1) still equals the number of negatively charged electrons (1).

4   They both have 8 protons / they have the same proton number / atomic number (1), and they both have 8 electrons (1) orbiting the nucleus. They have different numbers of neutrons: the first has 8 neutrons whereas the second has 10 (1). (both needed for second mark)

5   Any two of the following explanations:

   1) An atom can lose one or more electrons by friction (1) where contact forces rub electrons away (1) from the atom (1).

   2) An atom can lose one or more electrons by ionising radiation (1), where electrons are removed from the atom by an alpha or beta particle colliding (1) with an electron.

   3) An atom or molecule can lose one or more electrons by electrolysis (1) when it was previously bonded in an ionic compound and is separated in solution in an electrolytic cell (1).

37   Models of the atom

1   The plum pudding model showed the atom as a 'solid', positively charged (1) particle containing a distribution of negatively charged electrons (1) whereas the Rutherford model showed the atom as having a tiny, dense, positively charged nucleus (1) surrounded by orbiting negatively charged electrons (1).

2   Rutherford fired positively charged alpha particles at atoms of gold foil and most went through, showing that most of the atom was space / a void (1). Some were repelled or deflected (1), showing that the nucleus was positively charged (1).

3   (a) A (1)

   (b) The Bohr model showed that electrons orbited the atom at specific energy levels (1) and those electrons had to acquire precise amount of energy to move up to higher levels (1). The 'excited' electrons also emitted discrete amounts of energy to move to a lower level (1). The model was an improvement because it could explain emission and absorption spectra (1).

38   Radioactive decay

1   (a) Activity is the rate (1) at which the unstable / radioactive nuclei decay per second (1).

   (b) The unit of activity is the becquerel (Bq). (1)

   (c) Count rate is the number of counts of radioactive decay (1) per unit of time / second / minute. (1)

2   D (1)

3   In 1 second, 450 radioactive nuclei will decay (1) so $450 \times (2 \times 60)$ (1) = $54\,000$ (1) nuclei in 2 minutes.

4   (a) Alpha radiation / particle: (1) the alpha particle consists of 4 nucleons / 2 protons and 2 neutrons / a 'helium' nucleus (1)

   (b) Beta radiation / particle: (1) a neutron changes to a proton increasing the positive charge by 1 (1)

39   Nuclear radiation

1   B (1)

2   alpha – very low, stopped by 10 cm of air

   beta minus – low, stopped by thin aluminium

   gamma – very high, stopped by very thick lead

   all correct, (2) 2 correct (1)

3   (a) No change in relative atomic mass. (1)

   (b) High-energy electron emitted from the nucleus. (1)

   (c) Moderately ionising. (1)

4 Compared with other types of ionising radiation, the chance of collision with air particles at close range is high **(1)** because the alpha particles have a large positive charge / are massive compared with other types of radiation **(1)**. Once an alpha particle has collided with another particle it loses its energy **(1)**.

5 Alpha and beta particles and gamma waves lose their energy when they collide **(1)** with shielding atoms, causing them (instead of body atoms) to become ionised **(1)**. The denser the shielding material the greater the chance of collision **(1)** and subsequent reduction in energy of the harmful radiation **(1)**.

40 Uses of nuclear radiation

1 (a) Beta **(1)** radiation is used because alpha radiation / particles **(1)** would not pass through and gamma radiation / waves / rays **(1)** would pass too easily.

   (b) (i) The paper has become too thick. **(1)**

   (ii) The pressure on the rollers would be increased to make the paper thinner. **(1)**

2 They have high frequency / they carry large amounts of energy. **(1)**

3 (a) Alpha particles cannot pass through to the outside of the smoke alarm **(1)** and they are contained in a metal box / stopped by about 10 cm of air / are situated away from normal traffic of people **(1)**.

   (b) The smoke particles absorb the alpha particles **(1)** so the current falls / is broken and this triggers the bell to ring **(1)**.

4 Plastic instruments cannot always be heated to sterilise them **(1)** so gamma-rays can be used to kill bacteria/microbes **(1)**.

41 Nuclear equations

1 (a) α **(1)**

   (b) β− **(1)**

2 B **(1)**

3 (a) (for nitrogen) 7 **(1)**

   (b) (for magnesium) 12 **(1)**

4 (a) beta-plus (positron) **(1)**

   (b) alpha particle **(1)**

   (c) neutron **(1)**

5 (a) add 208 to Po **(1)**; type of decay = alpha **(1)**

   (b) add 86 to Rn **(1)**; type of decay = alpha **(1)**

   (c) add 42 to Ca **(1)**; type of decay = beta-minus **(1)**

   (d) add 9 to Be **(1)**; type of decay = neutron **(1)**

42 Half-life

1 Half-life is the time taken for half the nuclei in a radioactive isotope to decay. **(1)**

2 (a) 8 million atoms **(1)**

   (b) 9.3 minutes = 3 half-lives **(1)** so, after 1 half-life, 8 million nuclei left, after 2 half-lives, 4 million nuclei left, and, after 3 half-lives, 2 million nuclei left **(1)**

3 (a) The activity is 400 Bq at 1.5 minutes **(1)** (between 1.3 and 1.7 is allowed). Half this activity is 200 Bq, at 6.5 **(1)** minutes (between 6.3 and 6.7 is allowed), so the half-life is 6.5 min to 1.5 min = 5 min **(1)**. (Answers between 4.7 and 5.3 min are allowed.) *(If you used other points on your graph and got an answer of around 5 min you would get full marks. For this question your working can just be pairs of lines drawn on the graph.)*

   (b) Net decline tells you what ratio **(1)** of the radioactive material has decayed after each half-life **(1)**.

   (c) 7/8th of the radioactive material has decayed after three half-lives. **(1)**

43 Contamination and irradiation

1 Before 1920, the effects of radium were not known / recognised **(1)** so it was thought that it was safe to use **(1)**. It was banned from use once the dangers were known **(1)**.

2 (a) External contamination: radioactive particles come into contact with skin, hair or clothing. **(1)**

   (b) Internal contamination: a radioactive source is eaten, drunk or inhaled. **(1)**

   (c) Irradiation: a person becomes exposed to an external source of ionising radiation. **(1)**

3 (a) Any suitable example, e.g. contaminated soil may get on to hands. **(1)**

   (b) Any suitable example, e.g. contaminated dust or radon gas may be inhaled. **(1)**

4 Internal contamination means that the alpha particles come into contact with the body through inhalation or ingestion **(1)**, where they are likely to cause internal tissue damage **(1)**. Alpha particles that are irradiated are less likely to cause damage because they have to travel through air **(1)** and are therefore less likely to ionise body cells **(1)** (at distances of over 10 cm).

44 Hazards of radiation

1 A **(1)**

2 1. Limiting the time of exposure / keep the time that a person needs to be in contact with the ionising source as low as possible. **(1)**

   2. Wear protective clothing / wearing a lead apron will absorb much of the ionising radiation. **(1)**

   3. Increasing the distance from the radioactive source / the further a person is from the ionising radiation, the less damage it will do. **(1)**

3 A source of alpha particles with high activity inside the body will ionise body cells **(1)** because they are highly ionising / massive / undergo many collisions **(1)** before transferring all of their ionising energy. Gamma-rays can pass out of the body fairly easily **(1)** without causing much damage to cells **(1)**.

4 Radioactive tongs allow the source to be kept as far as possible away from a person's hand **(1)** and allows it to be pointed away from people at all times **(1)**.

5 Those who use X-rays on a regular basis, such as medical workers, would have a high exposure / dose of radiation, which would cause damage if the dose was too high **(1)**. They leave the room so that they are not exposed to high levels of cumulative radiation / high dose **(1)**. The number of X-rays that a patient has is carefully monitored to minimise risk of high dose / exposure to high radiation **(1)**.

45 Background radiation

1 Radon is a radioactive element **(1)** that is produced when uranium in rocks decays **(1)**.

2 Levels can vary because of the different rocks **(1)** that occur naturally in the environment. They can also vary due to the use of different rocks such as granite **(1)** in buildings.

3 Natural: air **(1)** and **two** of the following – cosmic rays, rocks in the ground, food **(1)**

   Manufactured: nuclear power, medical treatment, nuclear weapons **(1)** (all three needed for mark)

4 (a) (add the three values and divide by 3) SE 0.27 Bq, SW 0.30 Bq **(1)** (both needed for mark)

   (b) The South West (has the highest average level of background radiation) **(1)**

46 Medical uses

1 B **(1)**

2 Any **three** from the following: maximising the distance from the source of radiation; **(1)** using special tools such as tongs and gloves when handling the radiation **(1)**; minimising the time of exposure to the radiation **(1)**; shielding bodies from exposure to the radiation with thick concrete barriers **(1)**; shielding bodies from exposure to the radiation with thick lead plates / aprons **(1)**

3 A medical tracer is a radioactive solution that contains a gamma-emitting radioisotope **(1)**. It is injected into the patient and is then absorbed by the organ being examined **(1)**. A special camera detects the gamma radiation emitted by the solution **(1)**. The detected waves are used to build up an image of where the radioisotope is located in the organ **(1)**.

4 The radioactive isotope must have a half-life long enough to give a useful image **(1)**, but short enough so that its nuclei have mostly decayed after the image has been taken **(1)**.

47 Nuclear fission

1 Left nucleus: uranium-235 **(1)**. Two new nuclei: daughter nuclei **(1)**. Small single particles – neutrons **(1)**.

2 (a) Nuclear fission is the splitting **(1)** of a large and unstable nucleus (e.g. uranium or plutonium). **(1)**

   (b) Usually, for fission to occur the unstable nucleus **(1)** must first absorb a neutron **(1)**.

3 (a) A chain reaction occurs as neutrons are released from a fission reaction and are absorbed by more uranium nuclei that also undergo fission **(1)**. As more than one neutron is released per fission reaction **(1)** this leads to more and more fission reactions occurring which could lead to an explosion **(1)**.

Answers

(b) Control rods are used to absorb excess neutrons to keep the chain reaction at a steady rate. **(1)**

(c)

uranium-235   uranium-235   uranium-235

neutron   neutron   neutron   neutron
and so on

fission fission fission

neutron   neutron   neutron

neutron   neutron   neutron

Uranium atom absorbing / about to absorb a neutron **(1)**. Three neutrons and energy released **(1)**. At least two more uranium atoms shown **(1)** each releasing three neutrons **(1)**.

(d) Controlled chain reactions are used in nuclear power stations. **(1)**

48 Nuclear fusion

1 B **(1)**

2 Fusion needs high pressures and temperatures **(1)**. At present, this requires more energy than is released by fusion **(1)**.

3 When fusion occurs the mass of the products is slightly less than the mass of the reactants **(1)** so the difference is released as energy in the form of light and heat **(1)**.

4 (a) Nuclei need to get very close to each other before fusion can happen **(1)**. Electrostatic repulsion means that the positive charges of the nuclei repel each other **(1)**.

(b) To give the nuclei enough energy **(1)** to overcome this, very high temperatures **(1)** and pressures **(1)** are needed.

49 Extended response – Radioactivity

The answer should include some of the following points: **(6)**

- All three types of radiation can pass into / penetrate different materials.
- Alpha particles have high relative mass and so transfer a lot of energy when they collide, so they are good at ionising.
- Alpha particles produce a lot of ions in a short distance, losing energy each time.
- Alpha particles have a short penetration distance so are absorbed by low density / thin materials such as a few centimetres of air and a sheet of paper.
- Beta particles have a low relative mass and can pass into / through more materials than alpha particles.
- Beta particles are less ionising than alpha particles and can be absorbed by 3-mm-thick aluminium.
- Gamma waves are high-frequency EM waves and can travel a few kilometres in air.
- Gamma waves are weakly ionising and need thick lead or several metres of concrete to absorb them. **(6)**

50 Scalars and vectors

1 (a) Scalars: speed, energy, temperature, mass, distance **(1)**

Vectors: acceleration, displacement **(1)** force (or weight), velocity, momentum **(1)**

(b) Any correct choice and explanation, e.g. mass **(1)** is a scalar because it has a size / magnitude **(1)** but no direction **(1)**.

2 (a) (i) Velocity is used because both a size and a direction are given. **(1)**

(ii) The students are jogging in opposite directions so the negative sign for one student indicates this. **(1)**

(b) The length of the arrow is proportional to the magnitude of the vector (in this case velocity) **(1)**. An increase in velocity to 3 m/s would mean an arrow 1½ times longer than that for 2 m/s **(1)**.

3 (a) D **(1)**

(b) Weight has size / magnitude and direction but the other quantities just have a magnitude. **(1)**

51 Interacting forces

1 gravitational **(1)**, magnetic **(1)**, electrostatic **(1)**

2 A **(1)**

3 Weight and normal contact force are vectors because they have a direction **(1)**. Weight is measured downwards **(1)** whereas normal contact force is measured upwards/opposite to weight **(1)**.

4 (a) pull (by the student on the bag) and friction/drag of the bag against the floor **(1)**.

(b) weight and normal contact/reaction force **(1)**

5 As the skydiver leaves the plane, weight acting downwards is greater than air resistance acting upwards so he accelerates **(1)**. As speed increases, air resistance increases to become equal to weight, so there is no net force / terminal velocity reached **(1)**. When the skydiver opens the parachute, air resistance upwards is greater than weight downwards so he decelerates **(1)**. The skydiver decelerates until air resistance upwards equals weight downwards – there is no net force (so a new terminal velocity is reached) **(1)**. (or similar wording)

52 Gravity, weight and mass

1 (a) The mass of the LRV on the Moon is 210 kg **(1)** because the mass of an object does not change if nothing is added or removed. **(1)**

(b)

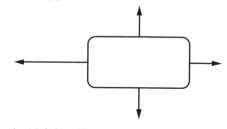

Arrow points vertically downwards / top of arrow estimated around the seat area of the LVR **(1)**. (both points needed for mark) The centre of mass is where the weight of a body can be assumed to act downwards through a single point **(1)**.

2 $W = m\,g$ **(1)** so (1 + 2 + 1.5) kg × 10 N/kg = 4.5 kg × 10 N/kg **(1)** = 45 N

3 Calculating correct masses for all three items. **(1)**

Selecting correct items **(1)** (clothes 10.5 kg + camera bag 5.5 kg + jacket 3.5 kg) = 19.5 kg **(1)**

53 Resultant forces

1 (a) A: 9.5 N; **(1)** B: 2 N; **(1)** C: 4.5 N; **(1)** D: 12.75 N **(1)**

(b) A: up; **(1)** B: up; **(1)** C: to the right; **(1)** D: to the left **(1)**

2 B **(1)**

3 (a) diagonal arrow – 2.5 cm **(1)**

(b) 50 N **(1)**

4 Scale correct, e.g. vertical 2 cm (6 N) and horizontal 5 cm (15 N) **(1)**

Hypotenuse = 5.4 **(1)** cm; represents 16.2 **(1)** N.

54 Free-body force diagrams

1 B **(1)**

2 (a) Upward **(1)** and downward **(1)** – same length.

(b) Force upwards 20 N **(1)**

Weight downwards 20 N **(1)**

3 (one only for all arrows) **(1)**

Longest arrow to left indicating movement to left. **(1)**

4 (a) 3.6 cm **(1)**

(b) 7.2 N **(1)**

55 Work and energy

1 A **(1)**

2 (a) Work done against friction will lead to a rise in temperature of the object **(1)** which is dissipated to the environment **(1)**.

(b) The greater the amount of friction, the more work that has to be done **(1)** to move the body through the same distance **(1)**.

3 $s = W / F$ so $s = 4800 / 80$ **(1)** = 60 **(1)** m

4 (a) $h = E / (m\,g)$ so $h = 320 / (8 \times 10)$ **(1)** (remember to convert the grams to kilograms) = 4 **(1)** m

(b) Use $W = F\,s$ so $F = W / s$ **(1)** and $F = 320 / 4$ **(1)** = 80 **(1)** N

56 Forces and elasticity

1 C **(1)**

2 (a) Tension: washing line (or any valid example) **(1)**

(b) Compression: G-clamp, pliers (or any valid example) **(1)**

(c) Elastic distortion: fishing rod (with a fish on the line) (or any valid example) **(1)**

(d) Inelastic distortion: dented can or deformed spring (or any valid example) **(1)**

3 After testing, Beam 1 would return to the same size and shape as before the test **(1)** and would be intact **(1)**. Beam 2 would distort and change shape **(1)** but would (probably) still be intact **(1)**.

4 Car manufactures install crumple zones / seat belts / air bags (1) in cars. These are parts of the car body that are designed to distort / change shape (1) in the event of a crash. They extend the time taken for a body to come to rest, reducing the force on the body (1).

57 Force and extension

1 Elastic deformation means that the object will change shape in direct proportion to the force(s) exerted, up to the limit of proportionality (1), and the change in shape is not permanent (1). Inelastic deformation means that the object will change shape beyond the limit of proportionality / the limit of proportionality is exceeded (1) and the change in shape will be permanent (1).

2 Extension = 0.07 m - 0.03 m = 0.04 m

Force = (spring constant / k) × extension = 80 N × 0.04 m (1)

Force = 3.2 (1) N (1)

3 D (1)

4 (a) $F = k e$ so $k = F / e$ (1) = 30 N / 0.15 m (1) = 200 (1) N/m

(b) $E = \frac{1}{2} k e^2 = \frac{1}{2} \times 200$ N/m × (0.15 m)^2 (1) = 2.25 J (1)

58 Forces and springs

1 (a) Hang a spring from a clamp attached to a retort stand and measure the length before any masses or weights are added using a half-metre ruler, marked in mm (1). Carefully add the first mass or weight and measure the total length of the extended spring (1). Unload the mass or weight and re-measure the spring to make sure that the original length has not changed (1). Add at least five masses or weights and repeat the measurements each time (1).

(b) The elastic potential energy can all be recovered (1) and is not transferred to cause a permanent change of shape in the spring (1).

(c) Masses must be converted to force (N) by using $W = m \times g / F = m \times g$ (1). The extension of the spring must be calculated for each force by taking away the original length of the spring from each reading (1). Extension measurements should be converted to metres. (1)

(d) (i) The area under the graph equals the work done/the energy stored in the spring as elastic potential energy. (1)

(ii) The gradient of the linear part of the force–extension graph gives the spring constant k. (1)

(e) Limit of proportionality. (1)

(f) energy stored = $\frac{1}{2} \times k \times e^2$ (1)

2 The length of a spring is measured with no force applied to the spring whereas the extension of a spring is the length of the spring measured under load/force less the original length. (1)

59 Moments

1 B (1)

2 When an object is balanced the clockwise moment (1) is equal to the anticlockwise moment (1).

3 $M = F d = 25$ N × 0.28 m (1) = 7 (1) N m (1)

4 (a) No (1)

Moment for Ben = 300 N × 0.8 m = 240 N m (1)

Moment for Amberley = 250 N × 1.2 m = 300 N m (1)

(b) Moment for Ben must change to equal 300 N m so distance must change to 300 N m / 300 N = 1 m (1)

(c) Original moment for Ben = 240 N m, so new moment for Amberley must equal this (1)

240 N m / 250 N = 0.96 m (1) (so Amberley must move to 0.96 m from the pivot)

60 Levers and gears

1 B (1)

2 (a) (i) input forces as shown (1) (both needed)

(ii) output forces as shown (1) (both needed)

(iii) pivots as shown (1) (both needed)

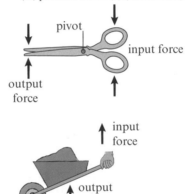

(b) The input force is the force provided by the user of the lever (1). The output force is the force that results from the input force (1).

3 For a high gear, the driver gear has a greater diameter than the driven gear (1) and the output is low for a given input force. (1) For a low gear, the driver gear has a similar diameter to the driven gear (1) and the output is high for a given input force. (1)

4 The cyclist would change to a low gear when moving from a horizontal road to a hill because a high output is needed (1) for a given force to move the bicycle up the hill against gravity (1).

61 Pressure and upthrust

1 D (1)

2 $P = F / A$ so $P = 25 / 0.0625$ (1) = 400 (1) Pa

3 The weight of the box is found by $F = P A$ so $F = 2000$ Pa × 0.25 m^2 (1) = 500 N (1)

4 (a) The more cargo a ship carries, the deeper it sits in the water (1). When ships are not carrying cargo, they weigh less and so displace less water (1) and the hull / Plimsoll line is higher above the surface of the water (1).

(b) Warm seawater produces slightly less upthrust than cold seawater (1), so ships will float lower down in warm water (1) and the Plimsoll line will be higher up the hull to avoid the ship becoming overloaded (1).

62 Pressure in a fluid

1 Water pressure increases with depth (1) because there is a greater weight of liquid with increasing depth (1).

2 (a) $P = h \rho g = 1500$ m × 1025 kg/m^3 × 10 N/kg (1) = 15 375 000 (1) Pa (1)

(b) Pressure at a point increases with the height of the column of liquid above that point (1) and with the density of the liquid (1).

3 (a) Atmospheric pressure describes a column of air, reaching from the Earth's surface to the top of the atmosphere (1), and covering one square metre of the Earth (1) containing a mass of air equal to 10 000 kg (1).

(b) $P = 100\,000$ Pa and $A = 1\ m^2$ so force (weight) = 100 000 Pa × 1 m^2 = 100 000 N (1)

$F = m g$, or $m = F / g$ (1)

so $m = 100\,000 / 10 = 10\,000$ kg (1)

63 Distance and displacement

1 C (1)

2 Distance does not involve a direction and so is a scalar quantity (1). Displacement involves both the distance an object moves and the direction that is has moved in from its starting point so it is a vector quantity (1).

3 (a) The circumference of the Ferris wheel is $2 \pi \times (15 / 2) = 47.1$ m. (1) The wheel completed three cycles so the total distance travelled was 3×47.1 (1) = 141.3 m (1). (accept 141 m)

(b) The final displacement was 0 m (1) as the girls returned to the starting point at the end of the ride. (1)

64 Speed and velocity

1 Average speed = 10 000 m / 2400 s = 4.17 (1) m/s

2 (a) Any **three** of the following: age, terrain, fitness, distance travelled. (1)

(b) (i) walking: 1.5 m/s (1)

(ii) running: 3 m/s (1)

(iii) cycling: 6 m/s (1)

3 The speed of the satellite can be constant because the distance being covered each second is constant (1), but the velocity changes constantly because direction of motion is constantly changing (1).

4 (a) $s = v t$, so distance = $4 \times (15 \times 60)$ (1) = 3600 (1) m.

(b) The term velocity is used because the boat has both a speed (1) and a direction. (1)

(c) At the finish of the journey by boat, the displacement of the team from the boathouse is 3600 m (1). As they return to the boathouse by bus, the displacement decreases (1) until they arrive back at the boathouse where the final displacement is 0 m (1).

65 Distance–time graphs

1 (a) (i) B (1)

    (ii) C (1)

(b) Evidence of attempt to calculate the gradient of the slope to find speed / use of change in distance divided by the change in time / speed = distance / time, (1) so s = 20 / 40 (1) = 0.5 (1) m/s.

(c) Walking (slowly) (1) as the average walking speed is 1.5 m/s (1).

(d) In part A, he travels 60 m in 40 s. Speed = distance / time (1) = 60 m / 40 s (1) = 1.5 (1) m/s. (You can use any part A of the graph to read off the distance and the time as the line is straight; you should always get the same speed.)

(e) Displacement is the length and direction of a straight line between the runner's home and the park (1), but the distance the runner ran may have included bends and corners on the path that the runner took (1).

66 Velocity–time graphs

1 (a) $a = \Delta v / t = 4 / 5$ (1) = 0.8 (1) m/s^2 (1)

2 (a) A (1)

(b) A right-angled triangle with a horizontal side and a vertical side that covers as much of the line as possible for precision. (1)

(c) Change in velocity = 30 - 0 m/s. Time taken for change = 5 - 0 s. Acceleration = $\frac{\text{change in velocity}}{\text{time taken}}$ or $a = \Delta v / t = 30$ m/s / 5 s (1) = 6 (1) m/s^2. (The triangle drawn may be different but the answer should be the same.)

(d) Area under line or distance = ½ × 5 s × 30 m/s (1) = 75 (1) m

67 Equations of motion

1 Attempt to estimate area under the graph / count squares (1) = approximately 21 squares or ~ 21 (1) m. (allow 20–22)

2 (a) $a = \Delta v / t$ (1) = (25 m/s − 15 m/s) / 8 s (1) = 1.25 (1) m/s^2

(b) $v^2 = u^2 + 2as$ = (25 m/s)^2 + 2(1.25 m/s^2 × 300 m) (1) = 1375 m/s^2 (1)

$v = \sqrt{1375}$ m/s = 37 m/s (1) (allow rounding error – answers between 37.00 m/s and 37.10 m/s)

(c) $v^2 − u^2 = 2 a s$, so $s = (v^2 − u^2) / (2 a)$ (1) = ((5 m/s)^2 − 1375 m/s^2) / (2 × −2 m/s^2) = (1) = −1350 m/s^2 / −4 m/s^2

Distance = 337.5 m (1)

68 Terminal velocity

1 C (1)

2 (a) The force of gravity pulls the skydiver downwards (1). Air resistance is small so the resultant force is large (1).

(b) The skydiver stops accelerating because the air resistance becomes equal to her weight (1) so the resultant force is zero (1).

(c) Terminal velocity (1)

(d) The larger surface area increases air resistance (1) so the resultant force acts upwards (1) slowing the skydiver down (1).

3 Ben should measure and mark the column with equal distances (e.g. every 20 cm) (1). The ball should then be dropped through the viscous liquid and the time recorded between each marker (1). When the ball passes through two marked distances at the same speed, terminal velocity is reached (1). Ben can then use $s = d / t$ to calculate the speed at which terminal velocity occurs (1).

69 Newton's first law

1 Four arrows drawn: vertical: down = weight, up = upthrust (arrows the same length); horizontal left to right = driving force from the engines, right to left = water resistance or resistive force; the driving force arrow should be longer than the resistive force arrow. (1 mark for all the forces correctly named and 1 mark for all the corresponding relative lengths of the arrows)

2 (a) Resultant force = 30 N + (−5 N) + (−1 N) (1) = 24 (1) N

(b) The resultant force is zero / 0 N (1) so the velocity is constant / stays the same (1).

3 (a) Assume downwards is positive, so resultant downward force is positive = 1700 N − 1900 N (1) = −200 (1) N. (You should state which direction you are using as the positive direction.)

(b) The velocity of the probe towards the Moon will decrease (1) because the force produces an upward acceleration / negative acceleration (1).

(c) After landing, the forces on the probe are balanced so there is zero resultant force / the probe does not move (1). The probe will not move unless it is acted on by another force (1). The tendency of a body / the probe to remain at rest is called inertia (1).

70 Newton's second law

1 (a) The trolley will accelerate (1) in the direction of the pull / force (1).

(b) The acceleration is smaller / lower (1) because the mass is larger (1).

2 (a) $F = m a = 3000$ kg × −3 m/s^2 (1) = −9000 (1) N

(b) Backwards / in the opposite direction to the motion of the minibus. (1)

3 (a) $a = F / m = 10\,500$ N / 640 kg (1) = 16.4 (1) m/s^2

(b) The mass of the car decreases, (1) so the acceleration will increase (1).

71 Force, mass and acceleration

1 Electronic equipment is much more accurate (1) than trying to obtain accurate values for distance and time to calculate velocity, then calculate acceleration (1), when using a ruler and a stopwatch. (Reference should be made to distance, time and velocity.)

2 Acceleration is inversely proportional to mass. (1)

3 Acceleration is the change in speed ÷ time taken so two velocity values are needed (1); the time difference between these readings (1) is used to obtain a value for the acceleration of the trolley.

4 (a) For a constant slope, as the mass increases, the acceleration will decrease (1) due to greater inertial mass (1).

(b) Newton's second law, $a = F \div m$ (1)

5 An accelerating mass of greater than a few hundred grams can be dangerous and may hurt somebody if it hits them at speed (1). Any two of the following precautions: do not use masses greater than a few hundred grams (1), wear eye protection (1), use electrically tested electronic equipment (1), avoid trailing electrical leads (1).

72 Newton's third law

1 D (1)

2 As the rocket sits on the launch pad, its weight downwards is equal and opposite to (1) the reaction force upwards of the launch pad (1), so the rocket does not fall through.

3 The weight of the penguin is pushing down on the ice and the reaction force of the ice is pushing back on the penguin (1), so the penguin is supported by the ice and does not fall through (1).

4 Newton's third law says that the force must be equal in size/magnitude (1) and opposite (1) in direction for equilibrium. The force exerted by the buttresses on the building (1) is equal and opposite to the force exerted on the buttresses by the building (1) resulting in no movement occurring.

73 Stopping distance

1 (a) Thinking distance + braking distance = overall stopping distance (1)

(b) Speed increases by 3 times so thinking distance increases by 3, and therefore thinking distance = 3 × 6 m = 18 m. (1) Speed increases by 3 times so braking distance increases by 9 times and braking distance = 9 × 6 m = 54 m. (1) Overall stopping distance = 18 m + 54 m = 72 m (1)

(c) Thinking distance will increase if: the car's speed increases, the driver is distracted, the driver is tired, or the driver has taken alcohol or drugs (1). (All four points needed for mark.)

Braking distance will increase if: the car's speed increases, the road is icy or wet, the brakes or tyres are worn, or the mass of the car is bigger (1). (All four points needed for mark.)

2 $F d = ½ m v^2$, so $F = (½ m v^2) / d$, so $F = [½ × 1500$ kg × (8 m/s × 8 m/s)] / 75 m (1) = (750 kg × 64 m/s) / 75 m (1) = 640 (1) N

3 Driving faster will increase thinking distance (1) and braking distance (1). If drivers do not increase their normal distance behind the vehicle in front accordingly, there is an increased risk of an accident / collision (1).

74 Reaction time

1 B (1)

2 Human reaction time is the time taken between a stimulus occurring and a response (1). It is related to how quickly the human brain can process information and react to it (1).

3 (a) A person waits with his index finger and thumb opened to a gap of about 8 cm (1). A metre ruler is held, by a partner,

so that it is vertical and exactly level with the person's finger and thumb / with the lowest numbers on the ruler by the person's thumb **(1)**. The ruler is dropped and then grasped by the other person as quickly as possible **(1)**.

(b) 0.2 s to 0.9 s **(1)**

(c) The distance measured on the ruler would be short for the person with a reaction time of 0.2 s **(1)** and longer for the person with a reaction time of 0.9 s **(1)**.

75 Momentum

1 The momentum of the car would change if it accelerates / speeds up or decelerates / slows down **(1)**. The momentum of the car would also change if it changed direction **(1)** because velocity is a vector **(1)**.

2 $p = m\,v$ so $p = 1200$ kg $\times$ 30 m/s **(1)**. momentum = 36 000 kg m/s **(1)** in the south / southerly direction **(1)**.

3 (a) Momentum of Dima and his car = 900 kg $\times$ 1.5 m/s **(1)** = 1350 **(1)** kg m/s

(b) (i) Momentum of Sam and car = 900 kg $\times$ 3 m/s = 2700 **(1)** kg m/s

(ii) The total momentum is conserved / total momentum of both cars is unchanged. **(1)**

(iii) Total momentum = 1350 + 2700 = 4050 kg m/s **(1)** so velocity after collision = total momentum / total mass = 4050 kg m/s / 1800 kg **(1)** = 2.25 m/s **(1)**

4 Momentum of skater 1 before collision = 50 kg $\times$ 7.2 m/s = 360 kg m/s; momentum of skater 2 before collision = 70 kg $\times$ 0 m/s = 0 kg m/s, so momentum of both skaters after collision = 360 + 0 = 360 kg m/s **(1)**, so combined velocity = 360 kg m/s / (70 + 50) kg **(1)** = 3 **(1)** m/s.

76 Momentum and force

1 Force is the rate of change of momentum. **(1)**

2 (a) $p = m\,v$ **(1)** = 1500 kg $\times$ 25 m/s **(1)** = 37 500 kg m/s **(1)**.

(b) $F = m\,\Delta v / \Delta t$ (to find Δv use answer from (a): Δv = 25 m/s) so F = 1500 kg $\times$ 25 m/s / 1.8 s **(1)** = 20 833 **(1)** N

3 (a) The forces exerted on the passenger are large when the mass **(1)** or the deceleration **(1)** of the vehicle are large.

(b) An air bag / crumple zone / seat belt **(1)** increases the time over which a passenger comes to rest so reducing the force exerted on them **(1)**.

4 $F = m\,\Delta v / \Delta t$ so F = 500 kg $\times$ 5 m/s / 20 s **(1)** = 125 **(1)** N

77 Extended response – Forces

The answer should include some of the following points: **(6)**

- Acceleration is the rate of change of velocity (speed in a given direction) so although the speed is constant, the direction is continually changing for an object in circular motion.
- For motion in a circle there must be a resultant force known as a centripetal force that acts towards the centre of the circle.
- The string provides the centripetal force which acts towards the centre of the circle.

- Extend the investigation with different lengths of string.
- Extend the investigation with different masses.
- Improve data collection with electronic sensors.
- Improve data collection with video analysis.
- Reference to the importance of control variables for valid data collection.

78 Waves

1 When energy travels through water, we can see that the water particles themselves do not travel. This is because an object on the surface will 'bob' up and down / vertically **(1)** as the wave passes horizontally **(1)**.

2 When energy travels through air, we can see that the sound is generated by a vibrating / oscillating surface **(1)** (e.g. a loudspeaker) that causes air particles to vibrate / oscillate in the same plane, creating pressure waves / areas of compression and rarefaction / sound waves **(1)**.

3 (a) B **(1)**

(b) 12 cm / 2 (as there are 2 waves shown) so wavelength = 6 cm = 0.06 m **(1)**

(c) Any correct wave with higher amplitude (height) **(1)** and shorter wavelength **(1)**.

(d) To find frequency use $v = f\lambda$, so f = 3 $\times$ 10^8 m/s / 0.12 m **(1)** = 2 $\times$ 10^9 Hz **(1)** (or 2 $\times$ 10^6 kHz / 2500 MHz / 2.5 GHz)

To find time period use $period = 1 / frequency$ so $period = 1 / 2 \times 10^9$ **(1)** (allow error carried forward from finding frequency) = 5 $\times$ 10^{-10} seconds / s **(1)**

79 Wave equation

1 (a) $v = f\lambda$ rearranged is $f = v / \lambda$, so f = 1500 m/s / 88 m **(1)** = 17 **(1)** Hz

(b) $\lambda = v / f$ so λ = 1500 m/s / 22 Hz **(1)** = 68.2 m **(1)**

2 Wave speed = 0.017 m $\times$ 20 000 Hz **(1)** = 340 **(1)** m/s

3 $\lambda = v / f$ = 0.05 m/s / 2 **(1)** = 0.025 **(1)** m **(1)**

4 First, calculate the velocity of the waves s = d / t **(1)** = 300 000 000 m/s (or 3 $\times$ 10^8 m/s) **(1)**, then use your calculated velocity to find the frequency using $v = f\lambda$, so $f = v / \lambda$ or f = $(3 \times 10^8) / 5$ **(1)** = 6 $\times$ 10^7 **(1)** Hz.

80 Measuring wave velocity

1 Frequency (f) = 3 Hz, wavelength (λ) = 0.05 m, speed of waves = 3 Hz $\times$ 0.05 m **(1)** = 0.15 **(1)** m/s **(1)**

2 D **(1)**

3 T = 4 divisions $\times$ 0.005 ms = 0.02 s **(1)**

$period = 1 / frequency$ so $period = 1 / 0.02$ s **(1)** = 50 Hz **(1)**

81 Waves in fluids

1 (a) Count the number of waves that pass a point each second and do this for one minute **(1)**; divide the total by 60 to get a more accurate value for the frequency of the water waves **(1)**.

(b) Use a stroboscope to 'freeze' the waves **(1)** and find their wavelength by using a ruler in the tank/on a projection **(1)**.

(c) wave speed = frequency $\times$ wavelength or $v = f \times \lambda$ **(1)**

(d) the depth of the water **(1)**

2 A ripple tank can be used to determine a value for the wavelength, frequency and wave speed of water waves, **(1)** as long as small wavelengths **(1)** and small frequencies are used **(1)**.

3 water: hazard – spills may cause slippages; safety measure – report and wipe up immediately **(1)**; electricity: hazard – may cause shock or trailing cables; safety measure – do not touch plugs/wires/switches with wet hands or keep cables tidy **(1)**; strobe lamp: hazard – flashing lights may cause dizziness or fits; safety measure – check that those present are not affected by flashing lights **(1)**

82 Waves and boundaries

1 (a) C **(1)**

(b) Stone is more dense **(1)** and, the greater the difference in density between materials, the more sound energy will be reflected **(1)**.

2 Light waves from an object (on the surface) are reflected at a plane surface **(1)**. The angle of incidence equals the angle of reflection **(1)** when measured against a normal drawn perpendicular to the plane surface **(1)**. (Both required for mark.) This is repeated at the bottom of the periscope so that the light reaches the user **(1)** (or similar wording).

3 (a) incident sound

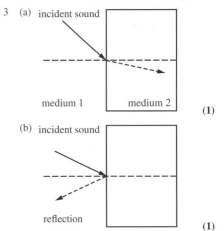

medium 1   medium 2 **(1)**

(b) incident sound

reflection **(1)**

4 Sound is mostly transmitted through the boundary from one material to another when their densities are similar **(1)** but some refraction will occur **(1)**. Sound is mostly reflected at the boundary between a low-density material and a relatively high-density material **(1)**.

83 Investigating refraction

1 (a) Place a refraction block on white paper and connect a ray box to an electricity supply **(1)**; switch on the ray box and set it at an angle to the surface of the block **(1)**; use a sharp pencil to draw around the refraction block and make dots down the centre of the rays either side of the block **(1)**; use a sharp pencil and ruler to join the 'external' rays and then draw a line across the outline of the block to join the lines **(1)**; use a protractor to draw a normal where the light ray met the block and measure the angle of incidence and angle of refraction **(1)**.

(b) When a light ray travels from air into a glass block, its direction changes **(1)** and the angle of refraction will be less than the angle of incidence **(1)**.

Answers

(c) (i) The ray of light would not change direction. **(1)**

   (ii) The light would slow down **(1)** (travelling from a less dense medium to a more dense medium) and the wave fronts would be closer together/ the wavelength of light would be shorter **(1)**.

2 any four from: use of electricity: if mains electricity is used there is a risk of shock – use tested apparatus/do not try to plug in/ unplug in the dark **(1)**; experiments are generally done in low-level light so there is a risk of tripping – clear floor area and working space (no trailing wires) and avoid moving around too much **(1)**; if glass blocks are used there is a risk of cuts – handle with care, use Perspex/non-breakable blocks when possible **(1)**; ray boxes get hot so risk of burns – do not touch ray boxes during operation **(1)**.

3 The waves travel more slowly and wavelength becomes shorter in shallow water **(1)**; the waves change direction/bend towards the 'normal' as they move into shallower water **(1)**.

84 Sound waves and the ear

1 B **(1)**

2 The particles are much further apart in air than in water **(1)** so it takes longer for vibrations to be passed from one particle to another **(1)**.

3 (a) A sound wave causes vibration of the eardrum tissue **(1)**. The vibration is transferred to the three 'solid' bones in the middle ear **(1)**. The vibration than passes through the bones (hammer, stirrup, anvil / causes vibrations that move through fluid in the cochlea / inner ear) **(1)**.

  (b) The vibrations in the cochlea / inner ear cause small hair cells to move **(1)** and transform them into electrical impulses which are sent to the brain **(1)**.

4 In the human ear, the eardrum will not vibrate **(1)** if the wave frequency is less than 20 Hz or more than 20 kHz **(1)**. If there is no vibration of the eardrum, no sound is heard. **(1)**

85 Uses of waves

1 C **(1)**

2 As ultrasound waves pass into the body, some waves are reflected each time they meet a layer of tissue **(1)** with a different density **(1)**. The scanner detects the echoes **(1)** and the computer uses the information to make a picture **(1)**.

3 (a) The Earth's mantle becomes denser with increasing depth **(1)**; the wave speed depends on the density **(1)**, which depends on the increasing pressure **(1)**.

  (b) $\lambda = v / f = 7$ m/s / 0.05 Hz **(1)** = 140 **(1)** m

4 Distance travelled by ultrasound = speed × time **(1)** = 8400 m/s × 0.5 × 10^{-9} s **(1)** = 4.2 × 10^{-6} m **(1)**

  Distance of layer below the surface = ½ × 4.2 × 10^{-6} = 2.1 × 10^{-6} **(1)** m

86 Electromagnetic spectrum

1 (a) All waves of the electromagnetic spectrum are transverse waves **(1)** and

they all travel at 3×10^8 m/s / the same speed in a vacuum **(1)**.

  (b) Electromagnetic waves all transfer energy from the source of the waves to an absorber. **(1)**

2 (a) A: X-rays **(1)**

   B: visible light **(1)**

   C: microwaves **(1)**

  (b) Frequency increases from radio to gamma waves **(1)**. The energy of the waves increases with frequency **(1)**, so gamma-rays have the most energy and radio waves the least **(1)**.

3 $v = f\lambda$, so $f = v / \lambda$ **(1)** = 3×10^8 m/s / 240 m **(1)** = 1.25×10^6 **(1)** Hz

87 Properties of electromagnetic waves

1 C **(1)**

2 (a) Reflection – waves bounce off a surface; **(1)** refraction – waves change speed and direction when passing from one material to another **(1)**; transmitted – electromagnetic waves are transmitted when they pass through a material **(1)**; absorbed – different electromagnetic waves are absorbed by different materials **(1)**.

  (b) Any **two** valid examples: e.g. reflection – light on a mirror **(1)**; refraction – light through water **(1)**; transmission – radio waves passing through the atmosphere from transmitter to receiver / X-rays absorbed by the atmosphere **(1)**.

3 (a) Microwaves are shorter in wavelength **(1)** and higher in frequency **(1)** than radio waves.

  (b) Microwaves sent from the transmitter are able to pass through the ionosphere **(1)** and are received and re-emitted by the receiver to the ground **(1)**. Radio waves are sent from the transmitter but are refracted by the ionosphere **(1)** and then reflected back to the receiver on the ground **(1)**.

4 Space-based telescopes are outside the Earth's atmosphere **(1)** so they can detect the whole range of electromagnetic waves **(1)** that are emitted by stars and galaxies **(1)**.

88 Infrared radiation

1 (a) any four from: Fill Leslie's cube with hot water at a known temperature (e.g. wait until it falls to 80 °C before taking temperature measurements) **(1)**. Measure the temperature at a distance (e.g. 10 cm) **(1)** from one of the four sides of Leslie's cube for a period of time (e.g. 5 minutes) **(1)**. Take regular readings (e.g. every 30 seconds) **(1)**. Repeat method for the other three sides **(1)**.

  (b) independent variable: sides of Leslie's cube; **(1)** dependent variable: temperature **(1)**

  (c) Any four from: starting temperature of the water, distance of heat sensor/thermometer from the cube, length of time, same number of readings taken, same intervals of time for each temperature reading, use the same thermometer or temperature

sensor. All four correct for **2 marks**; three correct for **1 mark**; two or fewer correct for **0 marks**.

2 (a) and (b) Any one hazard and method of minimising it from: Hot water in eyes can cause damage **(1)** – always wear eye protection **(1)**. Boiling water can cause scalds **(1)** – place the kettle close to cube and fill the cube in its place **(1)**. The cube can cause burns **(1)** – do not touch the cube until the temperature reading is low. **(1)** Water and electricity can result in a shock **(1)** – when using an electrical temperature sensor, keep well away from water **(1)**. Trailing wires can be a trip hazard **(1)** – avoid trailing leads/tuck leads out of the way **(1)**.

3 (a) The bungs would minimise thermal energy transferred from the flasks through evaporation. **(1)**

  (b) (i) Dull and black surfaces are the best emitters and best absorbers. **(1)**

    (ii) Shiny and light surfaces are the worst emitters and worst absorbers. **(1)**

89 Dangers and uses

1 (a) Infrared waves: (any **two** from) night vision goggles / security sensor / TV remote control / cooking / thermal imaging camera **(1)**

  (b) Ultraviolet waves: (any **two** from) disinfecting water / sterilising surgical / scientific instruments / entertainment lighting / security marking **(1)**

  (c) Gamma waves: (any **two** from) sterilising food / treating cancer / detection of cracks (pipes / aircraft, etc.) **(1)**

  (d) Communication / mobile phones / satellites **(1)**

2 (a) The oral X-ray delivers the least amount of radiation, 0.005 mSv, which is the same as 1 day's normal exposure to background radiation. The next higher dose is for a chest X-ray which delivers 0.1 mSv, equivalent to 10 days of background radiation, and the highest dose is for the lung cancer screening, which delivers 1.5 mSv, equivalent to 6 months' exposure to background radiation. (X-rays in the right order, **(1)** dose related to equivalent background radiation for all three, **(1)** data quoted from table for all three **(1)**))

  (b) If a person has too many X-rays in a certain time, particularly higher-dose X-rays, they may be more at risk of damage to body cells **(1)** if the X-rays are not carefully controlled to allow the body to recover from the high-energy doses **(1)**.

3 Radio waves can be produced by oscillations in electrical circuits **(1)**. Radio waves can also induce oscillations **(1)** in electrical circuits by creating an alternating current **(1)** with the same frequency as the radio waves, when they are absorbed **(1)**.

90 Lenses

1 A converging lens bends the rays of light towards each other **(1)**. A diverging lens bends the rays of light away from each other **(1)**.

2 (a) If the lens is thicker, the focal point will be closer to the lens / shorter **(1)**. If the lens is thinner, the focal point will be further away from the lens / longer **(1)**.

(b) The focal length is the distance from the lens to the principal focus. **(1)**

3 (a) Thick lens: mid-ray continues unchanged; top and bottom rays angle down steeply **(1)**, focal point relatively close to the lens **(1)**.

Thin lens: mid-ray continues unchanged; top and bottom rays angle down less steeply than for lens 1 **(1)**, focal point further from lens **(1)**.

(b) Convex lens **(1)**

4 (a) Three construction rays **(1)**, divergent rays extrapolated **(1)**, focal point correctly identified **(1)**, focal length correctly labelled **(1)**.

focal length (*f*)

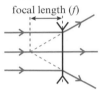

(b) (i) The image would be virtual **(1)**

(ii) Concave lenses only form virtual images **(1)** because the rays diverge / are extrapolated by the brain of the observer. **(1)**

91 Real and virtual images

1 A real image is an image that can be projected on a screen **(1)** but a virtual image cannot be projected on a screen **(1)**. A virtual image is produced when light rays appear to come from a point beyond the object **(1)**.

2 (a) Points added at the distance 2F on both sides of the lens. **(1)**

(b) Point added at the same distance as F on the right hand side of the lens. **(1)**

(c) Vertical arrow drawn from where the two rays of light cross on the right-hand side up to the central line. **(1)**

3 Image height = magnification × object height, so image height = 4.5 × 3.3 **(1)** = 14.85 mm **(1)**

4 (a) The magnification of the image increases. **(1)**

(b) When the object is closer to the lens than the focal point. **(1)**

(c) At 2F **(1)**

92 Visible light

1 C **(1)**

2 Any plane surface such as a mirror, very smooth water, polished glass. **(1)**

3 An opaque green object appears green because it reflects green light **(1)** and all other colours are absorbed **(1)**.

4 (a) Violet light has a shorter wavelength than red light **(1)** so it carries more energy.

(b) Yellow light is a combination of red light and green light **(1)**. The green book cover absorbs red wavelengths of light but reflects green wavelengths of light, so it still looks green **(1)**. The blue title absorbs both red and green wavelengths of light and so looks black **(1)**.

5

Incident rays of incidence all parallel, reflected rays reflected in different directions **(1)**. Uneven surface shown to represent rough surface **(1)**. Diffuse reflection still obeys the law of reflection because each ray of light that arrives at a surface is still reflected **(1)** according to the law of reflection (angle *i* = angle *r*) **(1)**. At the microscopic level, the surface is not smooth / the microscopic surfaces are at different angles to each other **(1)**.

93 Black body radiation

1 B **(1)**

2 Cup A will radiate more heat in 5 minutes **(1)** because the hotter the body the more infrared radiation it radiates in a given time **(1)**.

3 Maintenance of the average temperature of the Earth relies on a balance between absorption of energy from the Sun **(1)** and emission / radiation of energy into space **(1)**. Carbon dioxide absorbs thermal energy **(1)** and so reduces the amount of energy radiated into space. Scientists believe this leads to increased temperature / global warming of the Earth **(1)**.

4 Astronomers measure intensity of wavelengths of the light from a star to determine the peak / main wavelength **(1)**; they can then relate this to a scale of wavelength emitted at certain temperatures and so determine the temperature of the star **(1)**. (or similar wording)

94 Extended response – Waves

The answer should include some of the following points: **(6)**

- X-rays and gamma waves are both transverse waves.
- X-rays and gamma waves have a high frequency and therefore carry high amounts of energy.
- X-rays and gamma waves cause ionisation in atoms and exposure can be dangerous / cause cells to become cancerous.
- People who work regularly with X-rays and gamma waves should limit their exposure by using shields or leaving the room during use.
- X-rays are transmitted by normal body tissue but are absorbed by bones and other dense materials such as metals.
- X-rays and gamma waves mostly pass through body tissue but can be absorbed by some types of cells such as bone.
- X-rays and gamma waves can be used to investigate / treat medical problems.
- X-rays and gamma waves are used in industry to examine / 'see' into objects to examine for cracks / failures. **(6)**

95 Magnets and magnetic fields

1 Field lines **out** (arrows) of N **(1)**, field lines **in** (arrows) at S **(1)**, field line close at poles, **(1)** further apart at sides **(1)**.

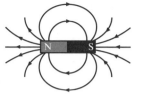

2 A bar magnet and the Earth both have north and south poles **(1)**. A bar magnet and the Earth have similar magnetic field patterns **(1)**. The direction of both fields can be found using a plotting compass **(1)**.

3 An induced magnet is used for an electric doorbell because it can be magnetised when the current is switched on **(1)**, which attracts the soft iron armature to ring the bell **(1)**, and is de-magnetised when the current is switched off **(1)** (moving the armature away from the bell). An induced magnet is used as the 'switching' off and on of the magnet allows the arm to be moved and the bell rung **(1)**. Also, the default 'rest' state of the bell does not need any energy input **(1)**.

96 Current and magnetism

1 (a) At least two concentric circles on each diagram. **(2)**

(b) Clockwise arrows on cross diagram **(1)**; anticlockwise arrows on dot diagram **(1)**.

2 B **(1)**

3 The force acting on a current-carrying wire in a magnetic field depends on the length / *l* **(1)** of the wire, the current / *I* **(1)** in the wire and the magnetic flux density / *B* **(1)**.

4 (a) $F = B\,I\,l$ so $F = 0.005 \times 1.4 \times 0.56$ **(1)** = 0.0039 **(1)** N

(b) $F = 0.005 \times 2.8 \times 0.56$ **(1)** = 0.0078 **(1)** N

(c) $F = 0.005 \times 1.4 \times 0.23$ **(1)** = 0.0016 **(1)** N

97 Current, magnetism and force

1 (a) When a current-carrying wire is placed in a magnetic field it would move **(1)**. This is because the current-carrying wire has its own magnetic field **(1)**, which interacts with the magnetic field / is repelled by the magnetic field / is attracted to the magnetic field **(1)**.

(b) D **(1)**

2 First finger – field **(1)**; second finger – current **(1)**; thumb – movement **(1)**.

3 The size of the force can be increased by increasing the strength of the magnetic field **(1)** or by increasing the current **(1)**.

4 $F = B\,I\,l$ so $I = F / Bl$ so $I = 0.21 \times 10^{-3}$ N / 0.0005 T × 0.30 m **(1)** so I = 1.4 **(1)** A **(1)**

98 The motor effect

1 Any one from: The wire can be coiled / turned into a solenoid **(1)**. The current can be increased. **(1)**

2 (a) Reverse the direction of current in the coil **(1)**. Reverse the direction of magnetic flux between the magnets **(1)**.

(b) Any valid example, e.g. washing machine / tumble dryer / airplane propeller / winch **(1)**

3 1. Increase the size of the current flowing in the coil. **(1)**

2. Increase the magnetic flux density between the poles of the magnet. **(1)**

3. Increase the number of turns on the motor coil. **(1)**

4 The commutator is in contact with the coil **(1)** but it is split into two parts, creating a gap **(1)**. As the motor spins, the contacts touch alternating sides of the commutator **(1)**, causing the current to change direction every half-turn **(1)**.

99 Induced potential

1 (a) Any three from: move the wire faster **(1)**, use a stronger magnetic field **(1)**, use more loops / turns in the wire **(1)**, wind the wire around an iron core **(1)**.

(b) The generator effect **(1)**

2 (a) When a wire or coil moves relative to a magnetic field **(1)** a potential difference is induced **(1)**, resulting in a current being generated. The conductor must cut across the magnetic field for a potential difference to be induced **(1)**.

(b) An induced current generates a magnetic field that opposes the original change. **(1)**

(c) An induced magnetic field may be due to the movement of the conductor **(1)** or a change in the magnetic field **(1)**.

3 (a) Sine wave sketched that goes above and below the x-axis. **(1)**

(b) The current moves between a positive value **(1)** and a negative value **(1)**.

100 Alternators and dynamos

1 Alternating current flows backwards and forwards / alternates in direction **(1)**. Direct current always flows one way / in the same direction **(1)**.

2 D **(1)**

3 (a) Graph A represents current produced by a dynamo. **(1)**

Graph B represents current produced by an alternator. **(1)**

(b) (i) Graph A shows that at every half turn / cycle the current does not change direction. **(1)**

(ii) Graph B shows that at every half turn / cycle the current changes direction. **(1)**

4 An alternator and a dynamo both use the interaction of conductors in magnetic fields **(1)** to produce an electric current **(1)**. As the coil in the alternator rotates, the way it faces is continually changing creating an alternating current through the slip rings **(1)** but, as the coil in a dynamo rotates, the way it faces also changes but, as the contacts also change on the split-ring **(1)** a direct current **(1)** is produced.

101 Loudspeakers

1 A **(1)**

2 A motor converts electrical energy into kinetic energy **(1)**. The loudspeaker does this by converting electrical energy into the movement of the cone **(1)**. This makes use of the magnetic property of electric current **(1)**, which interacts with the magnetic field of a permanent magnet **(1)**, causing a force.

3 The varying force exerted on the cone by the moving coil **(1)** results in movement / vibration of the cone **(1)**. The movement / vibration of the cone pushes the air causing pressure / sound waves **(1)**.

4 (a) A change in frequency of the vibration of the cone will result in a change of pitch in the sound. **(1)**

(b) A change in amplitude of the vibration of the cone will result in a change in volume / loudness of the sound. **(1)**

102 Transformers

1 A step-up transformer is used to increase potential difference and decrease current **(1)**. A step-down transformer is used to decrease the voltage and increase the current **(1)**.

2 When potential difference is increased the current is reduced **(1)**. The lower current **(1)** produces less heating effect **(1)** and therefore less energy is wasted in transmission of electricity **(1)**.

3 $V_p / V_s = n_p / n_s$, $V_p = 230$ V, $V_s = 19$ V, $n_s = 380$ turns; $n_p = V_p n_s / V_s$ **(1)** $= (230$ V $\times 380$ turns$) / 19$ V **(1)** $= 4600$ turns **(1)**

4 (a) This is a step-down transformer **(1)** because there are more turns on the primary coil than on the secondary coil **(1)**.

(b) $n_p = 600$ turns, $n_s = 20$ turns, $V_p = 360$ V, $V_s = (V_p n_s) / n_p$ **(1)** $= (360$ V $\times 20$ turns$) / 600$ turns **(1)** $= 12$ **(1)** V

103 Extended response – Magnetism and electromagnetism

The answer should include some of the following points: **(6)**

- A long straight conductor could be connected to a cell, an ammeter and a small resistor to prevent overheating in the conductor.
- When the current is switched on the direction of the magnetic field generated around a long straight conductor can be found using the right-hand grip rule.
- The right-hand grip rule points the thumb in the direction of conventional current and the direction of the fingers show the direction of the magnetic field.
- A card can be cut halfway through and placed at right angles to the long straight conductor. A plotting compass can be used to show the shape and direction of the magnetic field.
- The shape of the magnetic field around the long straight conductor will be circular/ concentric circles as the current flows through it.
- The strength of the magnetic field depends on the distance from the conductor.
- The concentric magnetic field lines mean that the field becomes weaker with increasing distance.
- The strength of the magnetic field can be increased by increasing the current.

104 The Solar System

1 Sun, planets, dwarf planets, moons (all required for mark) **(1)**

2 A planet orbits the Sun / a star **(1)** but a moon orbits a planet. **(1)**

3 Pluto was re-classified as a dwarf planet **(1)** because more powerful telescopes **(1)** have found many other objects similar to Pluto. **(1)**

4 (a) The Solar System has only one star (1) but the Milky Way contains over one billion stars. **(1)**

(b) The Solar System is a simple star system / has orbiting planets / is comparatively small **(1)** but the Milky Way is a spiral galaxy / collection of stars / contains planets **and** stars / is much bigger (accept any valid description of the Milky Way (accept any other valid difference). **(1)**

5 (a) 1. terrestrial / rocky planets **(1)**
2. gas giants. **(1)**

(b) C **(1)**

(c) Saturn is a gas giant whereas the other three planets are terrestrial / rocky planets. **(1)**

(d) The Earth has a solid crust with a separate gaseous atmosphere / atmosphere was produced through geological processes **(1)** whereas the surface of Neptune is gaseous and there seems to be no definite boundary between the 'surface' and the 'atmosphere'. **(1)**

105 The life cycle of stars

1 (nuclear) fusion **(1)**

2 B **(1)**

3 All stars exist for a time as main sequence stars but for massive stars this time is much shorter than for small stars **(1)**. After the main sequence stage, massive stars become red **supergiants** whereas smaller stars become red **giants** **(1)**. After the red supergiant stage, massive stars will explode as supernovae **(1)**, whereas smaller stars collapse to form white dwarfs **(1)**. Massive stars then go on to form either **neutron stars** or, for the most massive stars, **black holes** **(1)**. (Both points needed for final mark.)

4 All of the naturally occurring elements in the periodic table are produced by fusion processes in stars **(1)**. Elements heavier than iron **(1)** are produced in supernova explosions **(1)**. These explosions of massive stars **(1)** then distribute the elements throughout the Universe **(1)**.

106 Satellites and orbits

1 D **(1)**

2 The orbit of a planet is around a star (accept the Sun) **(1)** whereas the orbit of a moon is around a planet **(1)**.

3 (a) Planets, moons and artificial satellites all move in circular **(1)** orbits. (b) A moon orbits at a fixed distance **(1)** from its planet but the orbit of an artificial satellite can be changed by adjusting its speed **(1)** and the radius of its orbit **(1)**.

4 (a) A satellite is held in a circular orbit due to a balance **(1)** between the force provided by gravity **(1)** and its speed **(1)**. (b) A satellite in a stable orbit will move with changing velocity but unchanged speed because the satellite is continually changing direction **(1)** and velocity is a vector **(1)**. (c) If the speed of a satellite changes, the radius of its orbit **(1)** must also change **(1)**, to regain a stable orbit **(1)**.

107 Red-shift

1 D (1)

2 A galaxy with a red-shifted spectrum indicates that the galaxy is moving away from the observer (1). The further the black lines are shifted the faster it is moving away (1). As most galaxies are red shifted this would suggest that the Universe is expanding (1).

3 (a) The Big Bang theory. (1)

(b) The observation of supernovae (1) in red-shift (1).

(c) This suggests that the Universe began from one very small region (1) that was extremely dense (accept hot) (1).

(d) (i) Dark matter (1) and dark energy (1).

(ii) The Universe is believed to be expanding even faster than was first thought (1). There is much about the Universe that we do not yet understand (1).

108 Extended response – Space physics

The answer should include some of the following points: (6)

- The Big Bang theory states that the Universe started from a very small, hot and dense space in a massive explosion about 13.8 billion years ago.
- The Universe has been expanding and cooling ever since.
- The observed red-shift of light from galaxies provides evidence that the universe is expanding and supports the Big Bang Theory.
- There is an observed increase in the wavelength of light from most distant galaxies.
- The further away the galaxies are, the faster they are moving and the bigger the observed increase in wavelength. This effect is called red-shift.
- The observed red-shift provides evidence that space itself (the Universe) is expanding and supports the Big Bang theory.
- Scientists were able to either repeat the same experiment or test the theory using another method so the Big Bang theory was accepted.
- Since 1998 onwards, observations of supernovae suggest that distant galaxies are receding ever faster so this may lead to a theory that replaces the Big Bang theory.
- If future experiments find a better explanation of the origin of the Universe, then this theory will be adapted or replaced.

109 Timed Test 1

1 (a) (i) Background radiation (1)

(ii) Any two from: radon gas, (1) cosmic rays (1), medical uses (1), nuclear industry (1), natural sources/rocks (1)

(iii) Several readings are taken to identify anomalies to gain a reading close to the true value. (1)

(b) (i) y-axis labelled 'Corrected count rate (counts / min)' and x-axis labelled 'Time in minutes'; (1) all points plotted correctly (to within half a square) (2) [or six or more points correctly plotted (to within half a square) for 1 mark].

(ii) Single exponential decay curve of best fit passing through six points (1)

(iii) Half-life = 2 minutes (1)

(c) Alpha radiation is highly ionising because it is the most massive particle so it can easily ionise atoms by knocking off their electrons (1). Beta radiation is moderately ionising; the particles are highly energised although very small so have less chance of knocking electrons off other atoms (1). Gamma radiation is the least ionising because it has no mass but the gamma waves may still ionise other atoms (1).

2 (a) (i) $E_p = m g h$

Energy gained = 750 kg × 10 N/kg × 15 m (1) = 112 500 (1) J

(ii) Power $P = E_p / t$ = 112 500 J / 20 s (1) = 5625 (1) W (1)

(b) Any one of: thermal energy store of the motor (1), thermal energy store of the elevator materials (1), thermal energy store of the environment (1), sound energy store (1).

(c) $v = d / t$ = 15 m / 20 s (1) = 0.75 m/s (1)

Kinetic energy, $E_k = \frac{1}{2} m v^2 = \frac{1}{2} \times 750$ kg × (0.75 m/s)^2 (1) = 210.9 (1) J

(d) This process could be described as wasteful because it causes a rise in temperature in parts of the system so dissipating energy in heating the surroundings. (1)

(e) A rise in temperature in any one of: the lift motor (1), the fabric of the lift (1), the lift cables (1).

3 (a) (i) $P = V I$ / power = voltage × current (1)

(ii) $I = P / V$ = 2000 W / 230 V (1) = 8.7 A (1)

(iii) Too high a fuse could result in too much current and the risk of a fire (1); too low a fuse would melt and break the circuit each time the kettle was switched on (1).

(iv) C (1)

(b) 2000 W = 2000 J/s so E = 2000 J (1) and $E = Q V$, so $Q = E / V$ = 2000 J / 230 (1) V = 8.7 (1) coulombs / C

4 (a) Energy supplied $E = I V t$ = 12 A × 12 V × 120 s (1) = 17 280 (1) joules / J (1)

(b) $\Delta E = m c \Delta\theta$ so $\Delta\theta = \Delta E / (m c)$ = 17 280 J / (2 kg × 385 J/kg °C) (1)

Temperature change = 22.4 (1) degrees Celsius / °C (1)

(c) Some thermal energy is dissipated to the environment. (1)

(d) (i) By thermally insulating the block (1)

(ii) Insulator (1)

5 (a) (i) $^{231}_{91}\text{Pa} \rightarrow \, ^{227}_{89}\text{Ac} + \, ^{4}_{2}\text{He}$

i.e. 231, 89, 2 (1) (all three needed for mark)

(ii) Alpha decay (1)

(iii) $^{211}_{87}\text{Fr} \rightarrow \, ^{4}_{2}\text{He} + \, ^{207}_{85}\text{At}$

i.e. 211, 4, 85 (1) (all three needed for mark)

(iv) Alpha decay (1)

(v) $^{24}_{11}\text{Na} \rightarrow \, ^{24}_{12}\text{Mg} + \, ^{0}_{-1}\text{e}$

i.e. 12, 0 (1) (both needed for mark)

(vi) Beta decay (1)

(vii) $^{201}_{79}\text{Au} \rightarrow \, ^{201}_{80}\text{Hg} + \, ^{0}_{-1}\text{e}$

i.e. 79, 201 (1) (both needed for mark)

(viii) Beta decay (1)

(b) (i) Alpha decay causes both the mass and the charge of the nucleus to decrease. (1)

(ii) Beta decay does not cause the mass of the nucleus to change but does cause the charge of the nucleus to increase. (1)

6 (a) (i) As the fuel flows through the delivery pipe, electrons can build up due to friction (1) because the fuel pipe is made from an insulating material (1).

(ii) The build-up of static charge on an insulator could result in a spark (1). This would be dangerous because it could ignite the fuel / fumes / vapour (1).

(b) The further away the galaxies, the faster they are moving (1) and the bigger the observed increase in wavelength (1).

(c) *The answer should include some of the following points:* (6)

- A charged object creates an electric field around itself.
- The electric field is strongest close to the charged object.
- The further away from the charged object, the weaker the field.
- A second charged object placed in the field experiences a force.
- The force gets stronger as the distance between the objects decreases.

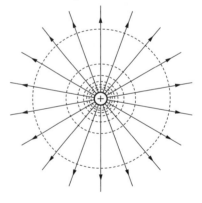

Diagram should include: positive point source, radial field lines, arrows outwards to show direction of field, equipotential lines, closer together nearer the centre.

7 (a) Each lamp would have a potential difference of 0.5 V (1) because potential difference is shared in a series circuit (1).

(b) Voltmeter connected in parallel across one of the lamps. (1)

(c) An additional cell could be added. (1)

(d) The lamps could be connected in a parallel circuit. (1)

(e) (i) The resistance of a thermistor changes with temperature (1). When the temperature rises more electrons are released (1), causing more current to flow and reducing resistance (accept converse) (1).

(ii) Used in a temperature-sensing circuit. **(1)**

8 (a) Solid – closely packed particles in fixed positions within a regular lattice.

Liquid – closely packed particles, irregular arrangement.

Gas – few particles, spread out in random arrangement (all three correct, **(2)** two correct **(1)**)

(b) *The answer should include some of the following points:* **(6)**

- Particles in a solid have a relatively low amount of kinetic energy compared with the other states of matter
   o and remain in molecular bonding with neighbouring atoms.
- Particles in a liquid have higher amounts of kinetic energy than solids
   o but lower amounts of kinetic energy than gases
   o and remain in weak molecular bonding with / can move around neighbouring atoms.
- Particles in a gas have a high amount of kinetic energy
   o and can move independently of other neighbouring atoms.

9 (a) $\rho_1 = m / V = 1.5$ kg / 0.2 m^3 = 7.5 kg/m^3 **(1)**

$\rho_2 = 0.7$ kg / 0.15 m^3 = 4.7 kg/m^3 **(1)**

$\rho_3 = 0.7$ kg / 0.2 m^3 = 3.5 kg/m^3 **(1)**

so block 1 has the highest density **(1)**

(b) (i) The student must make sure that the line of sight from the eye to the meniscus **(1)** is perpendicular to the scale **(1)** (to avoid parallax errors).

(ii) The student must measure the mass of the rock **(1)**. This can be done on a balance or hung from a suspended force meter **(1)**.

10 (a) Thompson's 'plum pudding' model referred to the atom as a positively charged sphere **(1)** with negative charges / electrons distributed throughout the sphere **(1)**. Rutherford's nuclear model of the atom showed the positive charges concentrated in a tiny nucleus **(1)** with the negative charges / electrons orbiting around the outside of the nucleus **(1)**.

(b) *The answer should include some of the following points:* **(6)**

- Alpha particles directed at gold foil
- Most alpha particles pass straight through
   o so most of the atom is empty space.
- A few alpha particles deflected through large angles
- Mass is concentrated at the centre of the atom
   o so the nucleus must be tiny.
- Nucleus is (positively) charged
   o so the negative charges orbit around the dense positive nucleus.

113 Timed Test 2

1 (a) Measure, mark and record the distance to be travelled by the trolley **(1)**, place the trolley at the start / top of the ramp and release **(1)**, time how long it takes to cover the marked distance **(1)** (both distance and timing should be mentioned for the mark). Repeat the experiment to reduce the influence of random errors **(1)**.

(b) Any two from: light gates **(1)**, data logger **(1)**, computer **(1)**

(c) (i) Independent variable: Height **(1)**

(ii) Column 2 and column 3: Distance, Time **(1)** (both needed for mark)

(iii) Speed (m/s) **(1)**

(d) (i) The forces acting opposite to the motion of the trolley / causing the trolley to stop would be friction between the wheels and the surface / floor / between the axles and wheels **(1)**

(ii) Friction could be reduced by (any one of the following): using a smoother surface on the ramp / smoother surface on the wheels / lubricating the axles **(1)**. Air resistance could be reduced by (any one of the following): making the trolley more aerodynamic / reducing the surface area of the front of the trolley **(1)**.

2 (a) $a = (v - u) / t$ **(1)**

(b) Average speed = 5 m/s, so $s = 5$ m/s × 20 s **(1)**

Distance = 100 m **(1)**

(c) $p = m v$ so momentum = 1000 kg × 10 m/s **(1)** = 10 000 **(1)** kg m/s **(1)**

(d) C **(1)**

3 (a) C **(1)**

(b) (i) Nuclear fusion is the joining of two light nuclei to form a heavier nucleus. **(1)**

(ii) Fusion processes in stars produce all the naturally occurring elements **(1)**. Fusion processes lead to the formation of new elements **(1)**.

(iii) Elements heavier than iron are produced in a supernova **(1)**. The explosion of a massive star (supernova) distributes the elements throughout the Universe **(1)**.

4 (a) (i) time period = 1 / frequency or $T = 1 / f$. **(1)**

(ii) 'T' should be marked between the start and end of an identified single wave cycle. **(1)**

(iii) First find T so T = 0.0001 / 2 (2 waves shown) so T = 0.00005 **(1)** so T = 1 / 0.000 05 s **(1)** = 20 000 **(1)** Hz.

(b) (i) Sonar **(1)**

(ii) The ship emits sound waves, which travel down to the shoal of fish **(1)**. Detectors on the ship receive the echo of the sound waves as they are reflected back from the fish **(1)**. The depth / location of the fish can be found by calculating the time between the sound wave being sent and the echo being detected **(1)**.

5 (a) Reflection **(1)**

(b) (i) i marked between incident ray and normal **(1)**, r marked between normal and refracted ray (inside the block) **(1)**.

(ii) As the wave passes from a less dense medium (air) to a denser medium (glass) the speed of light slows down **(1)** as the 'leading edge' of the wave front reaches the glass first. The wave then changes direction as the whole wave pivots on this 'leading edge' **(1)**.

(c) Rays drawn – tip of object to mid-lens then down through F$_2$ **(1)**. Second ray from tip of object through centre of lens to meet first ray **(1)**. Position of image shown **(1)**.

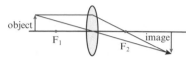

6 (a) A light source that is red-shifted will have a greater wavelength **(1)** and a lower frequency **(1)**.

(b) The further away the galaxies, the faster they are moving **(1)** and the bigger the observed increase in wavelength **(1)**.

(c) Red shift provides evidence of an expanding universe **(1)**, which supports the Big Bang model **(1)**.

(d) CMB radiation / cosmic microwave background radiation **(1)** which is detected from everywhere **(1)**.

(e) *The answer should include some of the following points:* **(6)**

- Scientists think that galaxies would rotate much faster if the stars they can see and detect were the only matter in galaxies. **(1)**
- Scientists have therefore proposed that there must be matter that cannot be detected or seen. **(1)**
- The concepts of dark mass **(1)** and dark energy **(1)** are not yet understood, **(1)** but one day may account for the 'missing matter' **(1)**.

7 (a) Planets and dwarf planets orbit round the Sun **(1)** but moons orbit round planets **(1)**.

(b) Jupiter is much larger than Mars / Jupiter is the largest planet in the Solar System **(1)**. Jupiter has many more moons than Mars / Jupiter has over 50 confirmed moons whereas Mars has 2 **(1)**.

(c) The circular path means a continual change of direction **(1)** and direction is a component of velocity **(1)**.

(d) *The answer should include some of the following points:* **(6)**

- At the start of a star's life cycle, the dust and gas / nebula are drawn together by gravity. **(1)**
- This causes fusion reactions. **(1)**
- The fusion reactions lead to an equilibrium between:
   o the gravitational collapse of a star **(1)** and
   o the expansion of a star due to fusion energy **(1)**.
- As the star loses mass, it expands to become a red giant. **(1)**
- The red giant then cools and collapses to become a white dwarf **(1)**.

8 (a) (i) The direction of the magnetic field
            will be clockwise **(1)**, as the current
            flows from positive to negative **(1)**
            (see diagram).

       (ii) The right-hand rule **(1)**

    (b) (i)

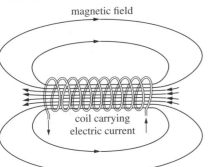

magnetic field

coil carrying
electric current

            Correct shape **(1)** lines of flux very
            close inside **(1)** lines of flux further
            apart outside, **(1)** pattern indicated on
            both sides of electromagnet. **(1)**

       (ii) Any valid use, e.g. doorbell, speaker,
            electromagnet **(1)**

    (c) $F = B\,I\,l = 0.5\text{ T} \times 3\text{ A} \times 0.75\text{ m}$ **(1)** $= 1.1$
         (1) N **(1)**

9 (a) The step-up transformer increases the
        potential difference leaving the power
        station so that the current is reduced
        (1). This reduces the amount of energy
        transferred to the thermal store / wasted
        during the transmission of electricity
        through the National Grid **(1)**.

    (b) $V_p / V_s = n_p / n_s$ so $V_s = (n_s\,V_p) / n_p$ **(1)** $=$
         $(4500 \times 250\text{ V}) / 150$ **(1)**

         Potential difference $= 7500$ **(1)** V

    (c) The high voltages that are necessary for
         efficient transmission via the National
         Grid are too dangerous **(1)** so a step-down
         transformer is used to reduce the voltage
         (1).

    (d) (i) $V_p\,I_p = V_s\,I_s$ so $I_s = (V_p\,I_p) / V_s$ **(1)** $=$
             $(4600\text{ V} \times 0.5\text{ A}) / 230\text{ V}$ **(1)**

             Current $= 100$ **(1)** A

       (ii) Assume that the transformer is 100%
             efficient. **(1)**

10 (a) The students have not unloaded the spring
        each time to check that the deformation is
        elastic. **(1)**

    (b) The students should use the same weights
         to get more reliable results to show a
         proportional relationship **(1)**; they should
         unload the spring after measuring each
         extension to check that the deformation is
         still elastic. **(1)**

    (c) 0.1:36, 0.2:40, 0.3:44, 0.5:52 (all four
         correct **(2)**; three correct **(1)**)

    (d) In linear elastic deformation, the
         extension (of a spring) is directly
         proportional to the force added **(1)**. The
         graph is a straight line which shows a
         directly proportional relationship **(1)**. The
         spring is loaded with weights of 0.1 N at
         a time **(1)** and the spring extends by 4 mm
         with each 0.1 N **(1)**.

Physics Equations Sheet

1	pressure due to a column of liquid = height of column × density of liquid × gravitational field strength (g)	$p = h \rho g$
2	(final velocity)^2 – (initial velocity)^2 = 2 × acceleration × distance	$v^2 - u^2 = 2 a s$
3	force = $\dfrac{\text{change in momentum}}{\text{time taken}}$	$F = \dfrac{m \, \Delta v}{\Delta t}$
4	elastic potential energy = 0.5 × spring constant × (extension)^2	$E_e = \dfrac{1}{2} k \, e^2$
5	change in thermal energy = mass × specific heat capacity × temperature change	$\Delta E = m \, c \, \Delta \theta$
6	period = $\dfrac{1}{\text{frequency}}$	
7	magnification = $\dfrac{\text{image height}}{\text{object height}}$	
8	force on a conductor (at right angles to a magnetic field) carrying a current = magnetic flux density × current × length	$F = B \, I \, l$
9	thermal energy for a change of state = mass × specific latent heat	$E = m \, L$
10	$\dfrac{\text{potential difference across primary coil}}{\text{potential difference across secondary coil}} = \dfrac{\text{number of turns in primary coil}}{\text{number of turns in secondary coil}}$	$V_p / V_s = n_p / n_s$
11	potential difference across primary coil × current in primary coil = potential difference across secondary coil × current in secondary coil	$V_s \, I_s = V_p \, I_p$
12	For gases: pressure × volume = constant	$p \, V = constant$